RISK AND SOCIAL THEORY IN ENVIRONMENTAL MANAGEMENT

RISK AND SOCIAL THEORY IN ENVIRONMENTAL MANAGEMENT

Thomas Measham
CSIRO Ecosystem Sciences

Stewart Lockie
Australian National University

CSIRO
PUBLISHING

National Library of Australia Cataloguing-in-Publication entry

Risk and social theory in environmental management/edited by Thomas Measham and Stewart Lockie.

9780643104129 (pbk.)
9780643104136 (epdf)
9780643104143 (epub)

Includes bibliographical references and index.

Environmental management – Australia.
Environmental risk assessment – Australia.
Risk – Sociological aspects.

Measham, Thomas B.
Lockie, Stewart.

363.70994

Published by
CSIRO PUBLISHING
36 Gardiner Road, Clayton VIC 3168
Private Bag 10, Clayton South VIC 3169
Australia

Telephone: [+613] 9545 8555
Local call: 1300 788 000 (Australia only)
Fax: +61 3 9662 7555
Email: csiropublishing@csiro.au
Web site: www.publishing.csiro.au

Front cover image by iStockphoto

Set in 10/12 Adobe Minion Pro and ITC Stone Sans
Edited by Adrienne de Kretser, Righting Writing
Cover and text design by James Kelly
Typeset by Desktop Concepts Pty Ltd, Melbourne
Index by Bruce Gillespie
Printed by Ingram Lightning Source

CSIRO PUBLISHING publishes and distributes scientific, technical and health science books and journals from Australia to a worldwide audience and conducts these activities autonomously from the research activities of the Commonwealth Scientific and Industrial Research Organisation (CSIRO). The views expressed in this publication are those of the author(s) and do not necessarily represent those of, and should not be attributed to, the publisher or CSIRO.

Feb26_RP_ILS

Contents

Preface

The idea for this volume initially evolved following a lively workshop of the Environment and Society Group attached to the Australian Sociological Association conference in December 2009. The group encouraged participation from a broad range of disciplines including human geography, political science and planning in addition to environmental sociology. This diversity was a defining characteristic of the papers presented, many of which have become chapters in this volume. Other chapters evolved from papers presented at the International Sociological Association World Congress in Gothenburg 2010, particularly an innovative session hosted by the Research Committee on Environment and Society (RC24).

The editors would like to thank Geoff Lawrence and Marcus Lane for helpful comments on developing the volume. We would also like to thank the reviewers who assisted with the peer review process including, in no particular order, Heinz Schandl, Andrew Reeson, Emma Yuen, Karin Hosking, Doug Cocks, Kirsten Maclean and Ben Harman in addition to the chapter authors who were able to assist with reviewing other chapters in the volume.

List of contributors

Cathy Banwell is a Fellow at the National Centre for Epidemiology and Population Health, Australian National University, with a background in social anthropology and public health. She specialises in the socio-cultural determinants of health risks associated with dangerous consumptions (food, alcohol, tobacco and illicit drug use).

Fabiana Barbi is a sociologist with a Master's degree in Environmental Science at the University of São Paulo (Brazil) and York University (Canada). She is a Research Fellow of the Earth System Governance Project and a PhD candidate in Environment and Society at the State University of Campinas, Brazil.

Leila da Costa Ferreira is full Professor at the Institute of Philosophy and Human Sciences and Centre of Environmental Studies and Research at the State University of Campinas, Brazil. She is a member of the editorial boards of *Journal Ambiente & Sociedade* and Annablume Press and a member of the International Sociological Association. She is Associate Faculty of Earth System Governance Project (International Human Dimensions Program) and was visiting Professor at Jiao Tong University, China in 2009.

Rafael D'Almeida Martins is a PhD candidate in Environment and Society at the State University of Campinas in Brazil and the coordinator of the Earth System Governance Research Fellows Network, Earth System Governance Project, a core project of the International Human Dimensions Program on Global Environmental Change.

Jane Dixon is a Senior Fellow and food sociologist at the National Centre for Epidemiology and Population Health, Australian National University. She has written extensively on the cultural and economic operations of supermarkets. She is currently undertaking research on the barriers to sustainable food production and consumption.

Tira Foran is a social scientist at CSIRO Ecosystem Sciences, Australia. His research background is in contentious politics and natural resource management. He has over 10 years of experience in environmental policy analysis, working in both the private sector and academia. His ongoing research interests involve how to improve planning and governance using techniques such as dialogue, participatory scenario-building and modelling.

Leonardo Freire de Mello has a PhD in Demography and works as a research collaborator at the Centre of Environmental Studies and Research, State University of Campinas, Brazil. He is Professor at the Paraíba Valley's University. His research focuses on the complex area of population–environment–consumption and the human dimensions of global environmental changes.

Thales Haddad Novaes de Andrade holds a PhD in Social Sciences at the State University of Campinas, Brazil and is Professor of Social Sciences at the Federal University of São Carlos. He is a Research Fellow of the National Council for Science and Technology Development, Brazil.

Fiona Haines is a criminologist and Associate Professor in the School of Social and Political Sciences at the University of Melbourne and Fellow of the College of Asia and the Pacific at the

Australian National University. Her expertise lies in the areas of globalisation, risk, regulatory theory and regulatory reform.

Brian Head leads the evidence-based policy and sustainability research programs in the Institute for Social Science Research, University of Queensland, Australia. His recent work includes urban water policy, environmental policy and governance, regional innovation and organisational aspects of sustainability including climate change and natural resources.

Andreas Klinke serves as Associate Professor in Environmental Policy in the Division of Social Science at the Environmental Policy Institute, Memorial University of Newfoundland, Canada. He received a PhD in Political Science at the Darmstadt University, Germany. His research and teaching activities include international environmental politics, risk governance, deliberation and participation in environmental policy.

Louis Lebel is the current and founding director of the Unit for Social and Environmental Research at the Faculty of Social Sciences, Chiang Mai University, Thailand. He has research interests in ecology, public health, development studies and environmental governance. He is active in global environmental change science programs.

Anne Leitch (BSc, GradDipComm, MComm) has a background in science communication. She is currently a PhD student working on issues of climate change and environmental planning and governance with the Centre of Excellence for Coral Reef Studies at James Cook University in Queensland, and CSIRO Ecosystem Sciences.

Stewart Lockie is Professor and Head of the School of Sociology at Australian National University, co-convenor of the National Institute for Rural and Regional Australia and president of the International Sociological Association's Research Committee on Environment and Society. Professor Lockie's research addresses environmental governance, sustainable agriculture, risk and safety management in hazardous industries and social impact assessment. Recent publications include *Agriculture, Biodiversity and Markets: Agroecology and Livelihoods in Comparative Perspective* (Earthscan Publications, 2010).

Alberto Matenhauer Urbinatti is an undergraduate student at the Institute of Philosophy and Human Sciences, Centre of Environmental Studies and Research, State University of Campinas, Brazil.

Tom Measham is a human geographer in the Social and Economic Sciences Program of CSIRO Ecosystem Sciences and Adjunct Associate Professor at the Fenner School of Environment and Society at Australian National University. His research interests are in social learning, community engagement and regional dimensions of sustainability science.

Lee M. Miller is Associate Professor of Sociology at Sam Houston State University (USA). Dr Miller studies community responses to disasters. Her research includes community reception of displaced people from Hurricanes Katrina and Rita in Walker County, Texas, the social control of disaster survivors following Hurricane Katrina and the delayed electric power restoration after Hurricane Ike. Current research projects focus on risk and community responses to perceived environmental threats.

Raymond Murphy is Emeritus Professor at the University of Ottawa, Canada. He is the author of *Sociological Theories of Education* (McGraw-Hill Ryerson, 1979), *Social Closure* (Oxford University Press, 1988), *Rationality and Nature* (Westview, 1994), *Sociology and Nature*

(Westview, 1997) and *Leadership in Disaster: Learning for a Future with Global Climate Change* (McGill-Queen's University Press, 2009).

Fernanda Oliveira de Souza is a biologist at the Institute of Biology, Centre of Environmental Studies and Research, State University of Campinas, Brazil.

Benjamin L. Preston is a senior research scientist at Oak Ridge National Laboratory (USA) and Deputy Director of Oak Ridge National Laboratory Climate Change Science Institute, where he conducts research on the societal impacts of climate change and the role of adaptation in reducing climate risk.

Steve Purbrick is a graduate engineer with Sinclair Knight Merz (Australia) working on water and environment issues. He completed a double degree in Civil Engineering and Arts (majoring in politics) at the University of Melbourne in 2008. He is currently a United Nations youth ambassador in Ghana.

Ortwin Renn is full Professor and Chair of Environmental Sociology and Technology Assessment at Stuttgart University, Germany. He directs the Interdisciplinary Research Unit for Risk Governance and Sustainable Technology Development at Stuttgart University and the non-profit company DIALOGIK, a research institute for the investigation of communication and participation processes in environmental policy making. He also serves as Adjunct Professor for Integrated Risk Analysis at Stavanger University, Norway and as Affiliate Professor at the Harbin Institute of Technology and Beijing Normal University.

Dr Cathy Robinson is a senior scientist who leads a Research Group at the CSIRO. Her research is focused on the critical analysis of indigenous and collaborative approaches to natural resource management programs in Australia and overseas.

Nick Schofield is a Senior Executive Consultant with Sinclair Knight Merz (Australia) specialising in water, climate, natural resource management, global issues and futures. He was formerly Science Manager for Land & Water Australia. He originally qualified with a PhD in Astrophysics but has spent most of his career in sustainability.

Bruce Taylor is a research scientist with the Social & Economic Sciences Program, CSIRO Ecosystem Sciences, Australia. His main research interests are in environmental governance in agricultural and urban landscapes, particularly in policy implementation settings that involve processes of regionalisation.

Tabatha Wallington is a research scientist with the Social and Economic Sciences Program at CSIRO, and Adjunct Senior Fellow with the School of Social Science at the University of Queensland. Her research is focused on knowledge systems and governance frameworks for environmental decision-making under conditions of uncertainty.

Rob White is Professor of Criminology in the School of Sociology and Social Work at the University of Tasmania. His latest book is *Transnational Environmental Crime: Toward an Eco-global Criminology* (Routledge, 2011).

1 SOCIAL PERSPECTIVES ON RISK AND UNCERTAINTY: RECONCILING THE SPECTACULAR AND THE MUNDANE

Stewart Lockie and Tom Measham

The idea of risk inevitably conjures images of the spectacular – the catastrophic failure of technological systems, the devastation of natural disasters, the creeping threat of global warming etc. While risk may be defined as any exposure to potentially negative consequences, it is usually only when the consequences are substantial that we begin, *en masse*, to take notice. And when risks are translated into realities – even ambiguous and contested realities – we especially begin to take notice. The concept of risk is not just an objective measure of potential harm, therefore, it is a cognitive and emotional bridge between the negative events affecting other people and our own fears and expectations. Take, for example, the 2011 Fukushima nuclear crisis. On 11 March, a magnitude 9 earthquake off the east coast of Japan triggered tsunami waves that swept up to 10 km inland, killing at least 15 000 people and damaging two reactors at the Fukushima nuclear power plant 250 km from Tokyo. Months later, efforts to contain radioactive fallout and repair the damaged reactors' cooling systems were ongoing. Within a few days of the earthquake that precipitated these events, media attention had started to shift from the humanitarian crisis and rescue efforts in devastated coastal towns to debate over the scale, causes and likely implications of damage to the Fukushima reactors. Long before a reliable picture began to emerge about the scale of that particular crisis, arguments between opponents and defenders of the nuclear industry became a major feature of the news cycle. These arguments were not solely, if ever, about Fukushima. They were about existing political agendas, protecting economic interests, reconciling different environmental imperatives and competing paradigms of risk calculation and management. They were also about the extent to which natural hazards may be amplified or alternatively mitigated by the technological and economic systems in which contemporary societies are embedded. Complex issues were boiled down by some into a straightforward (albeit false) choice between the risk of another Fukushima versus the risk of anthropogenic climate change.

It is easy to criticise protagonists in these debates, and the media which report them, for the unseemly haste with which conclusions were drawn about the implications of the Japanese tsunami for the rest of the world. At the same time, a disaster on this scale does demand critical analysis and debate both in and outside Japan. Social scientists will participate in this analysis, as they have in the aftermath of numerous other catastrophic events and as they do in relation to potentially catastrophic events associated with global warming. We will review social theory and research on risk and uncertainty in more detail in the following section. The point we wish to make here is that while this research converges around spectacular events and threats

(Kasperson *et al.* 1988; Wynne 1996), there is a pressing need to consider dimensions of risk that, on the surface, appear much more mundane. This is not a hard and fast distinction. Activities that seem routine today may be implicated in tomorrow's catastrophes. However, the smaller the perceived scale of risk, we contend, the less sustained sociological attention it is likely to attract. Risks that are localised, that affect a small number of people (or people with whom 'we' do not identify, such as the residents of developing countries), that are managed by small businesses or by local governments, or that have highly uncertain long-term impacts, have attracted comparatively limited interest from the social sciences. We are not seeking to sensationalise such risks by suggesting they are somehow commensurate with the 2011 events in Japan. Our intent is to expand the focus of social science research to better reflect the diversity of risk calculation and management activities in which environmental managers, both public and private, are engaged. Such activities range from the preparation of coastal management plans to industry policy, incentive design, risk communication and water conservation to corporate social responsibility programs and so on.

Social perspectives on risk and uncertainty

In popular discourse, the terms 'risk' and 'hazard' mean essentially the same thing – the possibility of being injured, harmed, or suffering some other negative consequence. Various professions and sciences ascribe more specific definitions that differentiate these terms. Specialised definitions are no more correct than common definitions, but they are useful in helping to bring clarity to our discussion of risk and in unpacking some of the theoretical assumptions and insights the social sciences, in particular, bring to bear on risk. A common conceptual clarification is the differentiation of 'risks' from 'hazards' on the basis of whether it is possible to quantify the probability of an undesirable event with any degree of confidence. In situations characterised by high levels of uncertainty, harmful events are generally classified as hazards while, in situations where probabilities are reasonably well known, they are classified as risks. Many situations may, of course, be characterised by competing calculations of risk, by both risks and hazards, by potential threats that are not easily classified as either a risk or a hazard and by threats that are not yet recognised and which may or may not be amenable to risk calculation. Luhmann (2002) proposed an important variation on the distinction between risks and hazards. He suggested that risks may be seen as the potential consequences that individuals or organisations understand enough to take into account in their own decision-making, and that hazards may be seen as the potential consequences of decisions taken by others, a process over which they have little or no control.

Numerous tools – many market-based – are used by businesses and governments to manage quantifiable risks (Matthews 2009). Quantification may be undertaken in numerous ways. Renn (2008a) offered a useful synthesis of different disciplinary approaches to risk measurement. Those drawn from the natural, economic and psychological sciences and professions are summarised in Table 1.1.

We do not intend to discuss in detail each disciplinary perspective summarised in Table 1.1. What we do wish to do is highlight the emphasis across these perspectives on reducing uncertainty and on encouraging rational responses to potential risk events based on their objective likelihood and magnitude. Where this is possible, there is no denying its desirability. One of the major emphases within social science perspectives on risk, however, has been on understanding the circumstances in which calculating and responding to risk probabilities in a straightforwardly technical manner is not possible. It is possible to identify a range of such situations.

Table 1.1: Non-social disciplinary perspectives on risk

Discipline	Risk concepts	Description	Applications
Natural and technical sciences	Actuarial analysis	Probability estimate of physical harm extrapolated from statistical data	Insurance
	Causal risk models	Quantify indirect and complex causal relationships between risk agents and consequences, e.g. dose–effect relationships between exposure to environmental pollutants and human health	Establishing regulatory threshold values for environmental toxins
	Probabilistic risk assessment	Estimate the probability and consequences of safety-relevant failures in complex technological systems. May address uncertainties through inclusion of confidence intervals informed by expert judgement	Improving safety performance in hazardous industries
Economics	Utility maximisation	Cost–benefit assessment of risk management options. Willingness to pay or revealed preference techniques may be used to integrate monetary and non-monetary values. Discount rates used to value future benefits may be varied to reflect the likelihood of risk events, uncertainty over their likelihood, ethical judgments etc.	Comparing decision options
Psychology	Risk perception	Attempt to understand the intervening variables and intuitive heuristics that consistently influence subjective perceptions of risk, e.g. degree of perceived personal control, extent to which risk is shared with others, familiarity with risk situation	Design of risk communication strategies

- *Hazards characterised by a high degree of uncertainty* (Pigeon 2008). This may simply be due to a lack of scientific knowledge on which to base risk calculation. However, uncertainty may also be due to scientific conflict and/or derive from complex interactions between human activities and ecosystem processes leading to non-linear and unpredictable trajectories of change, on the one hand, and/or social learning and adaptation on the other (Perez 2008). Anthropogenic climate change is a classic example. While the causal relationships between atmospheric greenhouse gas concentrations and temperature change are well understood, the specific magnitude and timing of climatic changes and their implications in discrete locales are subject to considerable uncertainty and controversy. Future technological and institutional responses that may mitigate/exacerbate or adapt/maladapt to climate changes are similarly unpredictable.
- *Residual uncertainties over the implications of otherwise well-known risk events.* Confidence in the likelihood or magnitude of a risk event does not necessarily translate into confidence at other scales (as discussed above) or into confidence over appropriate

management responses. As Lockie (Ch. 6, this volume) points out, for example, the tools already exist to quantify the probabilities of short- to medium-term climatic variability, but implementing appropriate management responses on the basis of probability-based predictions still poses a considerable practical challenge for agriculture and other natural resource-dependent industries.

- *Risk events that are extremely low in probability but catastrophic in consequence* (Pigeon 2008). Formal risk assessment cannot provide definitive guidance in such circumstances on whether the rational response is to ignore the risk on the basis of its remote probability or to avoid the risk at all costs due to its potential consequences. Cost–benefit analysis may assist by helping to factor in consideration of the relative risks and benefits of alternatives to the activity in question. In the end, though, there is no objective answer to questions such as whether we should forgo the benefits of an activity such as nuclear power generation in terms of economic development and greenhouse gas abatement due to the possibility of serious nuclear accidents.
- *Risk events that evoke value conflicts in relation to the acceptability of risk exposures and mitigation strategies* (Pigeon 2008). Debates over nuclear energy and the use of genetically modified organisms, for example, are focused as much on the ethics of particular technologies and the rights of those exposed to them to participate in decision-making as they are on the material probabilities of risk. What is at stake is not just the mathematical probability of a particular kind of harm but the type of society in which we wish to live and our willingness to trade off particular values in order to avert a threat or, conversely, maintain short-term benefits. Ambiguity concerning the goals and values that ought to be pursued through decision-making processes, and over who has rights to participate in that decision-making, are often misrepresented as questions of scientific uncertainty, a theme taken up by Taylor *et al.* (Ch. 12, this volume).
- *Dynamic relationships between risk assessment and risk management strategies* (Luhmann 2002). The relationships between risk identification, probability estimation and development of response mechanisms are not linear and sequential. Each of these activities is co-determined. Assessment of risk cannot be undertaken in isolation but must anticipate political, policy and management responses. The dynamic relationships between risk assessment and management therefore render them vulnerable to knowledge gaps concerning the different risks, risk perceptions and likely responses of actors across varying stakeholder groups (Luhmann 2002). For example, environmental scientists and economists might see a carbon trading scheme as an effective response to managing the known risks of increased atmospheric carbon dioxide, but overlook risks in other areas of society such as the political risks of introducing such a scheme, a topic taken up by Haines (Ch. 4, this volume).
- *Situations in which the costs and benefits that arise from risk-inducing activities, risk reduction or management activities and/or risk events themselves are unequally distributed* (Pigeon 2008). Fossil fuel-derived energy, for example, supports the bulk of contemporary economic activity and, by extension, employment, incomes, food security, welfare provision etc. At the same time, it is associated both with localised accidents and pollution and with global processes of atmospheric warming. Generally speaking, the 'winners' and 'losers' in this distribution of costs and benefits are separated socially, spatially and temporally. Frequently, those who stand to lose from risk-inducing activities have few options for meaningful participation in risk assessment or decision-making.
- *Situations characterised by a lack of confidence in the capacity or trustworthiness of expert and/or risk-regulating institutions* (Pigeon 2008). Public responses to numerous potential and actual risk events have been dismissed by key regulatory institutions

> either as irrational overreactions or as irrationally indifferent. Sociological studies, however, have identified several factors which suggest that divergence between lay and expert assessments of risk is not always irrational. For example, expert judgements may not accord with people's own experiences of risk consequences, institutions may have a history of systematically understimating the consequences of risk events, and institutions may systematically 'bracket out' and ignore important dimensions of the risk event and its consequences (Wynne 1996). This theme is taken up by Miller (Ch. 7, this volume). One dimension that is frequently bracketed out, for example, is the disproportionate exposure of some groups to threats such as pollution, toxic waste and food contamination. The environmental justice movement has shown that disproportionate exposure to risk is often concentrated in communities dominated by low socio-economic status and/or ethnic minority groups.

As Pigeon (2008) argued, it follows that simply providing more information or communicating more effectively will not resolve disputes between (and among) expert agencies and lay-people over risk. Useful typologies and overviews of more social scientific approaches to risk are offered by Pigeon (2008), Renn (2008a,b), Zinn and Taylor-Gooby (2006) and others. We will not summarise and critique every school of thought here. Rather, we will structure this introduction to social scientific approaches to risk around what we regard as major theoretical themes common to the various approaches, namely, the pervasiveness of risk, the construction and amplification of risk, and the need for more deliberative (or communicative) approaches to comprehending risk.

Beck (1992) famously argued that the distribution of exposure to risk is overtaking the distribution of wealth as the defining feature of social organisation in contemporary society. Modernisation is transforming us from the 'scarcity society' to the 'risk society'. It is true that long-standing political alliances and class conflicts continue to define much political debate, including environmental debate. We see evidence of this in research which shows that a significant proportion of US citizens now believe that scientific estimates of the risks posed by climate change are little more than politically motivated claims by the ideological left (McCright and Dunlap 2011). Despite these discontinuities, risks generated by industrial society – risks such as pollution, food contamination, nuclear fallout and climate change – have surpassed natural hazards in terms of their threat to human well-being (Beck 1992). In fact, these risks challenge any neat conceptual distinction between the 'natural' and the 'social' as they interact with, magnify and even subsume dangers we once might have attributed to the 'vagaries of nature' or rationalised as 'acts of God' (Beck 2010). Management of risks, as a consequence, is increasingly central to the functions and expectations of key social and political institutions. The relative invisibility of risks such as radioactivity, as well as numerous toxins and pollutants, intensifies dependence on those institutions, experts and media outlets capable of providing and interpreting information. However, at the same time that the proliferation of risks associated with industrialisation and modernisation demands the attention of scientific and bureaucratic institutions, the legitimacy of those institutions is called into question by their own role in the modernisation project (Beck 2010). Confrontation with the negative side effects of modernity thus has contradictory impacts and implications. The idea of 'reflexive modernisation' is used by Beck to describe a societal process through which the institutions in which scientific and technical expertise are embedded are simultaneously elevated in importance yet challenged by industry, citizen and non-government groups mobilising their own 'knowledge' and 'expertise'. In such a milieu, any claim to 'objectively' measure and manage risk must be treated with a degree of caution as all probability assessments depend upon normative judgements concerning the desirability of competing aspirations and interests along with what ought to be regarded as only tentative assumptions about the causal pathways between actions, risks and consequences.

Macro-social theory of the sort developed by Beck is always open to charges of overgeneralisation. However, numerous micro-sociological studies demonstrate that risk knowledge does not reside exclusively in scientific and technical institutions. Brian Wynne's (1996) well-known study of the interaction between scientific experts and sheep farmers involved in the management of radioactive fallout in the Cumbrian hills (UK) following the Chernobyl meltdown in 1986 illustrates this nicely. Cumbrian sheep farmers may well have depended on scientists for knowledge of the presence and behaviour of radioactive isotopes, but knew from their own experience that those same scientists made simplistic and often unrealistic assumptions about numerous aspects of the local agro-ecology (ranging from the distribution of soil types to the availability of feed and the dietary preferences of sheep) which also shaped the movement of radioactivity through the landscape. Scientific agencies not only ignored farmers' specialist knowledge of local environmental conditions, they denigrated farmers' suggestions, undermined trust and made numerous, ultimately false, pronouncements about how long sheep sales from the area would be affected. Wynne (1996) stressed that this conflict was not just a matter of competing claims about the technical 'facts' of risk. If it were, conflict could be resolved simply by paying more heed to local knowledge then integrating it rather more carefully with expert knowledge. Conflict, however, was also a matter of competing claims about normative issues such as who should be involved in the interpretation of 'facts' and how people ought to respond to them. Technical risk assessment, Wynne (1996) argued, is a form of cultural and moral intervention.

When social scientists describe risk knowledge as a social construct, therefore, they are not suggesting that such knowledge is an entirely subjective or individualised phenomenon but that knowledge about complex issues invariably includes symbolic and normative dimensions, and that risk claims are filtered through the interaction of existing frameworks of meaning, the social networks in which people are embedded, media and other communication channels, and so on. Filtering contributes to the contested and dynamic nature of dominant conceptualisations of risk and to the amplification or attenuation of specific risk claims in public discourse. The magnitude, distribution and time-frame of risks, trust or mistrust in claims-making institutions, and the complexity of cause and effect relationships, may all play a role in the amplification or attenuation of an issue (Pigeon 2008). Risks may also be amplified, both materially and symbolically, by the ripple effect of second- and third-order consequences. It follows that the ways in which people interpret and respond to risk information are situation- and context-specific (Pigeon 2008).

The perhaps unsurprising corollary of the argument that risks are far more social in nature than is implied by a straightforward calculus of probability and magnitude, is the argument that practices for calculating and managing risks should be similarly socialised. It is possible to make this argument on normative grounds (i.e. that people have a democratic right to involvement in decisions that affect their lives and livelihoods in major ways). However, it is also possible to make this argument on scientific grounds (i.e. that involvement of those affected by major risk scenarios is critical to the empirical validity of risk assessment and management). Risks and their impacts, in other words, are materially shaped through social and political processes.

More comprehensive research into public perceptions of risk and improved communication between expert and laypeople are important, but not by themselves sufficient, measures to address the social and political shaping of material risks. Social research again brings us back to the idea of improved participation in decision-making among those affected by risk through more deliberative approaches to risk assessment and management (Renn 2006; Renn and Schweizer 2009). The goal, informed by Habermas' (1984) concept of communicative action, is to provide forums that bring people and institutions together in order to share and evaluate knowledge, appreciate the interests, values and aspirations of different stakeholders, generate common understanding, highlight areas of uncertainty and ambiguity, and agree on actions (Renn 2006). Deliberative approaches are based on the assumption that while it is not always possible to satisfy all participants' goals and aspirations, it is possible for people's views 'to be swayed by rational arguments and to lay aside particular interests and opinions in deference to overall fairness and the common interest of the collectivity' (Miller 1992, p. 56). Given the complexity and fluidity of many risk situations, the role of deliberation is not to identify the single 'correct' risk calculation or management strategy but to arrive at decisions that participants believe are fair and reasonable (Miller 1992). This is a process of social learning as much as it is a process of negotiation and compromise.

The challenge lies in developing practical models through meaningful deliberation, particularly when risk situations involve large numbers of stakeholders or when stakeholders are hard to identify. Unfortunately, such models are few and far between in the literature. One exception, the cooperative discourse model (Webler *et al.* 1995), offers a practical three-step means of doing so within the framework of environmental impact assessment:

- identifying and selecting concerns and evaluative criteria;
- identifying and measuring impacts related to different policy options;
- conducting a discourse with randomly selected citizens acting as 'jurors' and representatives of various interest groups acting as 'witnesses' (Renn 2006).

This model provides a rigorous framework through which to evaluate and integrate various knowledge claims and values but, as a consequence of its use of representative rather than open processes of participation, it may struggle to convince many stakeholders of the legitimacy of its compromises and recommendations. Clearly, no one model will fit all situations. A variety of approaches to deliberation within risk assessment and management that are informed by, and contribute to, social theory and research are needed.

From the spectacular to the mundane

According to Beck (1992), the focus of most risk discourse on the spectacular allows for a strange and dysfunctional mutual complicity. All are responsible for the potentially catastrophic risks confronting humanity and therefore no individual producer, firm or government needs to accept responsibility. We see this repeatedly in climate change politics, with national governments and large corporations arguing repeatedly that even if anthropogenic climate change is 'real', no action to address it can be effective without universal cooperation. While one country or one firm continues business-as-usual with respect to greenhouse gas emissions, they argue, unilateral action to address our own emissions would be counterproductive (Bulkeley 2001). Despite this observation, there is a relative paucity of sociological risk literature that does not focus on the spectacular. Social research and theory has been biased, perhaps understandably, towards large-scale industrial hazards, the implications of these risks for macro-social processes of change and the conflicts that emerge when these risks are imposed on unwilling citizens (Kasperson *et al.* 1988; Wynne 1996). As the social science literature has moved to a lower level of resolution it has done so, therefore, with a focus on how people experience and perceive risk when confronted with scientific controversy and unacknowledged hazards. Relatively little literature at this level has addressed the experiences of environmental managers, particularly as they deal with risks that have not been amplified to the realm of the spectacular.

In this volume, we have attempted to re-balance the social science literature through a collection of chapters which complement existing foci by considering broader approaches to the social dynamics of risk, for example considering the motivations and constraints of state environmental managers, local government staff and social groups who have been marginalised in the dominant risk management discourses. The book is characterised by three defining features: empirical richness, broad disciplinary approaches to the social dynamics of risk, and a breadth of environmental issues and responses to dealing with risk at local and regional scales where risks are confronted by individuals and organisations. By focusing on empirical detail in these contexts – the first defining feature – the chapters collectively focus on the ways that environmental risks are framed, discussed and addressed beyond the popular discussion of the spectacular. In this way, the chapters move beyond conventional accounts of environmental pollution in icons such as the Australian Great Barrier Reef, through detailed discussion of the concepts, instruments and applications of risk governance. In a similar way, risks relating to climate change are deepened through discussions of the political risks for government decision-makers and the attempts of municipal planners to respond to flooding and heat stress. Through scrutinising empirical data, it becomes clear, for example, that ambiguity over questions of legitimacy and acceptability has been represented conveniently (but falsely) as a matter of scientific uncertainty. It also becomes clear that the role of context is vital. It is clear that what is considered a risk, let alone an acceptable level of that risk, varies across time and place.

The chapters draw on a range of disciplines, notably sociology, human geography, political science and planning theory, in addition to environmental research we can group under the title of sustainability science. Herein lays the second defining characteristic of this volume, namely a broad suite of approaches to the social dynamics of environmental risk. By this we mean that the book moves beyond conventional approaches to calculating the probability of predetermined variables. The chapters consider different types of risk, some of which can be readily quantified and others which can't, such as political risk and the risk of unethical outcomes or reputational damage. Issues relating to gender and ethnicity are recognised as social dimensions of risk. These dimensions reflect the ways that risks are perceived (Ch. 7) and how outcomes may be experienced (Ch. 15). In considering these social dimensions, it is important to note patterns of environmental victimisation (Ch. 3).

The third defining feature of this book is the wide breadth of environmental issues under consideration. The book includes classic issues such as non-point source pollution in agriculture, and current issues of concern relating to climate change. It considers energy security and the social and environmental risks associated with hydro power in large social and ecological systems such as the Mekong. The scope includes waste management relating to extractive industries, and potable water management as it relates to water supply and flooding. The book also includes issues relating to sustainable consumption and food systems (Ch. 13), to complement more conventional concerns with risks in agriculture.

In the next section we offer some broad insights for addressing social dimensions of environmental risk and the policy and implementation challenges entailed.

A new verse on risk: learning from social research on risk and environmental management

The social dimensions of risk go to the heart of environmental challenges, tensions and conflicts. Environmental issues, from global warming to food and energy security, are intrinsically related to risk. At the same time, each of these global issues manifests itself as a local challenge (Adger and Kelly 1999; Cutter *et al.* 2000; Naess *et al.* 2005). When it comes to dealing with environmental risk and uncertainty, the local scale is where much of the action happens, where the concepts and tools of addressing risk need to be applied and put to the test

(Mukheibir and Ziervogel 2007; Preston *et al.* 2009). This volume extends the social science of risk to take more seriously the biophysical embeddedness of human society. The chapters vary greatly in the theoretical concepts they discuss and in the empirical detail they present. Yet, across the chapters three core implications are present. The first is that scale and context matter. The second is the need to think broadly and the third, particularly relevant for academic audiences, is to focus on the practicalities of dealing with risk.

Scale and context matter

As noted above, confidence in the magnitude of risk at one spatial or temporal scale does not necessarily translate into a similar level of confidence at other scales. This may be due both to the particular scalar properties of the risk in question and to the mediation of risk/hazard dynamics by other factors such as local institutions, resources and ecosystems, interactions with other threats and so on. The ways that actors such as local governments, community groups and state agencies respond to environmental risks are influenced by a suite of context-specific factors. The first of these is the specific ways that hazards manifest at the local scale. We see, for example, that addressing the risk of injecting wastewater into the ground in Texas is confounded by the local legacy of defunct extraction wells in close proximity to human settlements (Ch. 7, this volume). The Texas case study also emphasises the importance of local discourses, which are characterised by social phenomena including gender and regional identity. Across the Pacific in Byron Bay, local residents attempt to manage a dynamic coastal zone in the best interests of all in the face of local pressures and uncertainties over the potential impacts of sea level rise (Ch. 9, this volume). These themes illustrate the importance of understanding local dynamics when confronted with the real-world complexity of developing appropriate risk management strategies (Fischer 2000).

When confronting complex problems such as climate change, the role of subnational institutions becomes crucial (Dovers and Hezri 2010). For example, in São Paulo there is a distinct set of political–institutional structures that define avenues for addressing the effects of climate change. Local institutions include secretariats, departments and directories and environmental legislation that may be mobilised in future climate change actions (Ch. 10, this volume). In the Mekong, asserting the importance of local context is central to improving the legitimacy of decisions taken at the national scale. Local livelihoods and legitimate local interests can easily be overlooked when the assessment of environmental risks is restricted to the national level (Ch. 15, this volume). Local resource endowments are also critical. In both Sydney and São Paulo, financial capacity and human resources (in addition to institutional design) were found to present distinct limits to response at the local scale and to raise important questions about cross-scale linkages (Ch. 11, this volume). It is crucial to consider how local communities can access the resources necessary to address risk in a more socially informed and integrated way. The resourcing of local institutions should not be taken for granted when seeking a deliberative approach to understanding risk.

Think broadly

Confronting risk at the local and regional scales requires flexibility in terms of theoretical and methodological approaches. An openness to alternative practices and perspectives is necessary to create the space for innovative reframing of problems to allow more robust responses in the face of change (Ch. 14, this volume). Conventional environmental risk assessment may be relevant in some cases, particularly those relating to spectacular icons such as the Great Barrier Reef, but there is no basis to assume it will always work in complex environmental circumstances (Ch. 8, this volume). Indeed, the premise of conventional risk assessment may inherently undermine community engagement because its focus relates to ascertaining the acceptable level of risk and determining how to keep that risk within defined economic or ecological parameters, without necessarily reflecting the diversity and dynamism of community concerns (Ch. 3, this volume).

Market-based instruments are often espoused in contemporary policy and academic discourses as appropriate tools to address a wide array of social and environmental problems (Ch. 6, this volume). These mechanisms can and do play a role in managing risk, but are of limited benefit when risk calculation is confounded by multiple unknowns. When uncertainty is high, a more fluid response is needed (Renn and Schweizer 2009). Deliberation on what is at stake, and definition of appropriate ways to respond, are key roles in complex risks which are not easily reduced to conventional assessments. Complex risks require a framework and culture of organisational learning in a context that is conducive to resolving cognitive, evaluative and normative conflicts over risk (Ch. 5, this volume). The use of scenarios in multi-stakeholder processes has proven a powerful way to resolve such conflicts and to consider the future consequences of decisions which may not capture all risks in conventional assessments (Kok *et al.* 2007). For example, Chapter 15 demonstrates how scenarios can improve the legitimacy of outcomes by giving voice to a wider set of stakeholders. It reminds us that it is important to consider who defines the risks of any given decision, and who is outside the decision-making process (Luhmann 2002).

The consequences of underestimating the risk of natural hazards emphasise the need to clearly recognise risk models, both formal and informal, as partial and provisional attempts to understand the material realities to which they refer (Ch. 2, this volume). Bearing in mind that such models also shape our responses to threats, and thus the level of risk we face, the relationships between risk calculations and risk exposures must be seen as recursive and co-constitutive. However, the role of risk interpretations and responses, along with local knowledge of risk dynamics, in shaping the material nature and magnitude of risk is not always accounted for in environmental management. The social sciences need to work more closely with the natural sciences to anchor risk mitigation both socially and biophysically in order to foresee and avoid harm.

Be practical

The chapters in this volume show that there is a distinct role for social science in addressing the practical implications of environmental risks. Yet beyond social scientists, theoretical concepts don't always translate easily into direct applications in concrete situations. An important message for scholars, students and applied researchers is that, when working with local governments and state agencies which are confronting risk in tangible ways, the relevance of theories of reflexive modernisation and so on may not be immediately apparent. The challenge for social scientists is to make their work accessible. In addition to their theoretical contribution, Taylor *et al.* (Ch. 12, this volume) demonstrate the practical problems for policy which occur when conflating technical uncertainty with moral ambiguity, and the importance of crafting politically and socially acceptable compliance outcomes.

The process of considering who is affected by risk and in what ways can be a crucial part of formulating a response to risk. Hence, social scientists have a valuable role. To perform it, they need to step away from the tendency to critique from afar and step into the context where risks are addressed in the social domain, where they can be debated and acted upon. This does not mean eschewing theory. Moving beyond the spectacular to the mundane, means much more than simply applying conventional risk assessment at a lower level of analytical or spatial resolution – several chapters have thus been influenced to varying degrees by the application of deliberative theory to managing risk. The processes through which risks are framed, discussed and acted upon can favour one type of knowledge over another. They can empower or disempower, include or exclude. Renn and Klinke (Ch. 5, this volume) show that a key criterion for the successful implementation of deliberative risk assessment is epistemological discourse, i.e. an exploration among participants not just of *what* they know, but *how* and *why* they know it.

It follows from the social science research on risk presented in this book that risk assessment and management need to consider:

- spatial and temporal variability and the different stakeholders, hazard/risk dynamics, values etc. implicated at different scales;
- local and other lay knowledges of hazard/risk dynamics and consequences;
- the distribution of hazard/risk exposure among interest groups;
- the values and value conflicts likely to be affected by a hazard/risk event or its management;
- the potential for (mis)trust in risk management institutions to amplify or attenuate both the symbolic and material dimensions of risk;
- the need for multiple interdependent actors to commit to risk management strategies even where these undermine short-term self-interest;
- the capacity of institutional arrangements for risk calculation and management to foster co-learning and adaptation.

References

Adger WN and Kelly PM (1999) Social vulnerability to climate change and the architecture of entitlements. *Mitigation and Adaption Strategies for Global Change* **4**, 253–266.

Beck U (1992) *Risk Society: Towards a New Modernity.* Sage, London.

Beck U (2010) Climate for change, or how to create a green modernity? *Theory, Culture and Society* **27**, 254–266.

Bulkeley H (2001) Governing climate change: the politics of risk society. *Transactions of the Institute of British Geographers* **26**, 430–447.

Cutter SL, Mitchell JT and Scott MS (2000) Revealing the vulnerability of people and places: a case study of Georgetown County, South Carolina. *Annals of the Association of American Geographers* **90**, 713–737.

Da Costa Ferreira L, D'Almeida Martins R, Barbi F, Matenhauer Urbinatti A, Oliveira de Souza F, Haddad Novaes de Andrade T and Freire de Mello L (2012) Risk and climate change in Brazilian coastal cities. In *Risk and Social Theory in Environmental Management.* (Eds TG Measham and S Lockie) pp. 133–145. CSIRO Publishing, Melbourne.

Dixon J and Banwell C (2012) Choice editing for the environment: managing corporate risks. In *Risk and Social Theory in Environmental Management.* (Eds TG Measham and S Lockie) pp. 175–184. CSIRO Publishing, Melbourne.

Dovers SR and Hezri AA (2010) Institutions and policy processes: the means to the ends of adaptation. *Wiley Interdisciplinary Reviews: Climate Change* **1**, 212–231.

Fischer F (2000) *Citizens, Experts and the Environment: The Politics of Local Knowledge.* Duke University Press, Durham, NC.

Foran T and Lebel L (2012) Using holistic scenarios to rewrite rural futures. In *Risk and Social Theory in Environmental Management.* (Eds TG Measham and S Lockie) pp. 199–216. CSIRO Publishing, Melbourne.

Habermas J (1984) *The Theory of Communicative Action. Vol. 1: Reason and the Rationalisation of Society.* Polity Press, Cambridge.

Haines F (2012) The lure of the market in tackling global warming. In *Risk and Social Theory in Environmental Management.* (Eds TG Measham and S Lockie) pp. 41–51. CSIRO Publishing, Melbourne.

Kasperson RE, Renn O, Slovic P, Brown HS, Emel J, Goble R, Kasperson JX and Ratick S (1988) The social amplification of risk: a conceptual framework. *Risk Analysis* **8**, 177–187.

Kok K, Biggs R and Zurek M (2007) Methods for developing multiscale participatory scenarios: insights from southern Africa and Europe. *Ecology and Society* **13**, 8.

Leitch AM and Robinson CJ (2012) Shifting sands: uncertainty and a local community response to sea level rise policy in Australia. In *Risk and Social Theory in Environmental Management.* (Eds TG Measham and S Lockie) pp. 117–131. CSIRO Publishing, Melbourne.

Lockie S (2012) Market-based resource management policy and environmental uncertainty: outsourcing risk calculation. In *Risk and Social Theory in Environmental Management.* (Eds TG Measham and S Lockie) pp. 79–90. CSIRO Publishing, Melbourne.

Luhmann N (2002) *Risk: A Sociological Theory.* 2nd edn. Transaction Publishers, New Brunswick.

Matthews M (2009) Fostering creativity and innovation in cooperative federalism: the uncertainty and risk dimensions. In *Critical Reflections on Australian Public Policy: Selected Essays.* (Ed J Wanna) pp. 59–70. ANU Press, Canberra.

McCright A and Dunlap R (2011) The politicization of climate change and polarization in the American public's views of global warming, 2011–2010. *Sociological Quarterly* **52**, 155–194.

Measham TG and Preston BL (2012) Vulnerability analysis, risk and deliberation: a case study of the Sydney Climate Change Adaptation Initiative. In *Risk and Social Theory in Environmental Management.* (Eds TG Measham and S Lockie) pp. 147–157. CSIRO Publishing, Melbourne.

Miller D (1992) Deliberative democracy and social choice. *Political Studies* **40**, 56–67.

Miller LM (2012) Women and risk: commercial wastewater injection wells and gendered perceptions of risk. In *Risk and Social Theory in Environmental Management.* (Eds TG Measham and S Lockie) pp. 91–104. CSIRO Publishing, Melbourne.

Mukheibir P and Ziervogel G (2007) Developing a municipal adaptation plan (MAP) for climate change: the city of Cape Town. *Environment and Urbanization* **19**, 143–158.

Murphy R (2012) Managing risk under uncertainty. In *Risk and Social Theory in Environmental Management.* (Eds TG Measham and S Lockie) pp. 17–26. CSIRO Publishing, Melbourne.

Naess LO, Bang G, Eriksen S and Vevatne J (2005) Institutional adaptation to climate change: flood responses at the municipal level in Norway: adaptation to climate change: perspectives across scales. *Global Environmental Change Part A* **15**, 125–138.

Perez P (2008) Embracing social uncertainties with complex systems science. In *Uncertainty and Risk: Multidisciplinary Perspectives.* (Eds G Bammer and M Smithson) pp. 147–155. Earthscan, London.

Pigeon N (2008) Risk, uncertainty and social controversy: from risk perception and communication to public engagement. In *Uncertainty and Risk: Multidisciplinary Perspectives.* (Eds G Bammer and M Smithson) pp. 349–361. Earthscan, London.

Preston BL, Brooke C, Measham TG, Smith TF and Gorddard R (2009) Igniting change in local government: lessons learned from a bushfire vulnerability assessment. *Mitigation and Adaptation Strategies for Global Change* **14**, 251–283.

Purbrick S and Schofield N (2012) Development of environmental risk management plans in Great Barrier Reef catchments. In *Risk and Social Theory in Environmental Management.* (Eds TG Measham and S Lockie) pp. 105–115. CSIRO Publishing, Melbourne.

Renn O (2006) Participatory processes for designing environmental policies. *Land Use Policy* **23**, 34–43.

Renn O (2008a) Concepts of risk: an interdisciplinary approach. Part 1: Disciplinary risk concepts. *GAIA* **17**, 50–66.

Renn O (2008b) Concepts of risk: an interdisciplinary approach. Part 2: Integrative approaches. *GAIA* **17**, 196–204.

Renn O and Klinke A (2012) Complexity, uncertainty and ambiguity in inclusive risk governance. In *Risk and Social Theory in Environmental Management.* (Eds TG Measham and S Lockie) pp. 59–76. CSIRO Publishing, Melbourne.

Renn O and Schweizer P-J (2009) Inclusive risk governance: concepts and application to environmental policy making. *Environmental Policy and Governance* **19**, 174–185.

Taylor BM, Wallington TJ and Robinson CJ (2012) Uncertainty and ambiguity in environmental governance: water quality in Great Barrier Reef catchments. In *Risk and Social Theory in Environmental Management.* (Eds TG Measham and S Lockie) pp. 161–173. CSIRO Publishing, Melbourne.

Wallington T, Robinson CJ and Head BW (2012) Crisis, change and water institutions in south-east Queensland: strategies for an integrated approach. In *Risk and Social Theory in Environmental Management.* (Eds TG Measham and S Lockie) pp. 185–198. CSIRO Publishing, Melbourne.

Webler T, Kastenholz H and Renn O (1995) Public participation in impact assessment: a social learning perspective. *Environmental Impact Assessment Review* **15**, 443–463.

White R (2012) Risk and environmental victimisation. In *Risk and Social Theory in Environmental Management.* (Eds TG Measham and S Lockie) pp. 27–39. CSIRO Publishing, Melbourne.

Wynne B (1996) May the sheep safely graze? A reflexive view of the expert–lay knowledge divide. In *Risk, Environment and Modernity: Towards a New Ecology.* (Eds S Lash, B Szerszynski and B Wynne) pp. 44–83. Sage, London.

Zinn JO and Taylor-Gooby P (2006) Risk as an interdisciplinary research area. In Risk in Social Science. (Eds P Taylor-Gooby and J Zinn) pp. 20–53. Oxford University Press, Oxford.

Part I

What does social theory tell us about environmental risk and vulnerability?

2 MANAGING RISK UNDER UNCERTAINTY

Raymond Murphy

In their study entitled *Catastrophe in the Making: The Engineering of Katrina and the Disasters of Tomorrow*, Freudenburg *et al.* (2009, pp. 165–166) concluded that 'there is no evidence, unfortunately, that our ability to foresee harm has kept up with our ability to create it'. This conclusion applies widely to harm caused by technologies that have resulted in the unleashing of deepwater oil gushers, destruction of fisheries, deforestation, habitat and diversity loss and so on. This chapter begins by suggesting some conceptual distinctions concerning risk and uncertainty necessary to foresee harm. It then examines the capacity to foresee harm when extreme weather struck vulnerable modern societies, and when there is an increasing capacity to create harm in terms of global warming. The objective is to contribute to a better understanding of risk and its management under conditions of uncertainty.

Necessary conceptual distinctions

A rigorous analysis requires precise concepts. If a concept conflates significantly different phenomena, then it leads to an ambiguous, confused and potentially misleading analysis. This is certainly true concerning studies of risk. What is risk? What is being referred to by expressions such as 'risk society', the 'social construction of risk' and the 'amplification (or attenuation) of risk'? Do they refer to real, objective, material dangers? After all, this is the way 'risk' is understood by the population, and it is important to avoid jargonising widely used terms and thereby creating artificial barriers to communication with the population. Unfortunately, some social theory has muddied the waters by using 'risk' to refer only to discourse about dangers, to what is talked about as threats and to perceptions of hazard (Hannigan 1995, 2006). This simplifies investigations because it circumvents the difficult task of determining whether risk perception is based on material dangers or on bogeymen, on impending peril or on paranoia, but it renders studies superficial by failing to take into account the threatening or safe biophysical context and hence the fallibility of perceptions of risk or safety.

Risk is a complex concept that refers both to what is valued and to material threats (Short 1989; Renn 2008). Hence its referent is influenced by culture and by biophysical dynamics. It refers to both social constructions and nature's constructions. I will use the standard definition of risk, 'the possibility that an undesirable state of reality (adverse effects) may occur as a result of natural events or human activities (Renn 2008, p. 98). It refers to the state of the material world, in the case examined here the biophysical world including human bodies, and to what is desired in that material world. Since risk refers to a future occurrence and is the opposite of safety, events eventually confirm the assessment of risk (harm from a hurricane striking New Orleans) or contradict it (the millennium bug failed to produce predicted dire consequences).

The severity of the consequences depends not only on the strength of the biophysical hazard but also on the socially constructed vulnerability of human communities. There are different types of risk, with an important distinction between routine risks (vehicle accidents) and extraordinary risks (earthquakes). Risk can be the consequence of a construction of nature's dynamics not resulting from human activities (a volcano) or the product of a social construction (a nuclear reactor). Risk needs to be kept conceptually distinct from awareness of risk, definition of risk, assessment of risk, measurement of risk and discourse about risk in order to investigate the relationship between those mental models and their biophysical referents. This is a crucial distinction precisely because the relationship is mediated by culture, habitus and power such that it is variable rather than risk determining awareness of risk. Risk can be denied or ignored but, far from eliminating risk, this increases the possibility of adverse effects. Note that I have revised my earlier terminology concerning 'perceived' or 'unperceived' risk to 'awareness' or 'unawareness' of risk because the risk of, for example, a tsunami or earthquake is not directly perceived by the senses in the way the tsunami waves or earthquake tremors are.

Risk awareness, assessment, measurement and discourse, on the other hand, refer to the state of knowledge or ignorance about harm and the material world. There is a significant difference between lay (everyday) risk assessment and scientific risk assessment. There are variants of the former, ranging from the lay knowledge of underprivileged groups to the unscientific and sometimes anti-scientific assumptions of powerful groups. The distinction between risk and risk awareness forces us to keep in mind that not only everyday but also scientific risk assessments are fallible. Scientific calculations of probabilities are more sophisticated and hopefully more accurate than everyday risk assessments, and in some cases they are the only possible risk assessment (ozone-layer depletion), but they too can be wrong. It is important to avoid mistaking scientific risk measurement for risk itself. Such calculations can be at variance with material risk, as occurred when NASA calculated it was safe to launch the Challenger space shuttle (Vaughan 1996).

Knowledge of risk can be high, as with routine and frequently occurring risks, or negligible, as with rare high-impact occurrences and under new conditions. The latter two are usually referred to as uncertainty and the threat is referred to as uncertain risk (Renn 2008). Ignorance is a key component of uncertain risks, an important variant being what Gross (2003) refers to as 'non-knowledge', i.e. we know that we do not know. Even scientific risk assessments are characterised by different degrees and types of uncertainty. They can be suggestive of risk but not be definitive, i.e. the knowledge upon which they are based is partial. Studies of the movement of tectonic plates lead to scientific assessments of the heightened risk of earthquakes, but the impact, location, timing, scope and duration of tremors remain unpredictable. Higher risk is known without knowing precisely what it will be. That knowledge is attained only when the event occurs. The same is true for hurricanes and global warming. The eventual arrival of danger can be foreseen, but uncertainties remain concerning the specifics. This illustrates that different kinds of uncertainty must be distinguished:

- total ignorance, where there is no good evidence about what the chosen path of development will lead to;
- partial knowledge, as in the earthquake and global warming examples where risk assessors know that there is increased risk without knowing precisely what it will be in terms of impact, timing, location, duration and who will be affected;
- erroneous knowledge, where lay or scientific risk assessors think they know the risks but are eventually proven wrong.

Disasters often demonstrate the limitations of pre-existing risk assessments and provide new information to learn lessons and improve assessments. This occurs in an incremental fashion surprise after surprise, disaster after disaster. There can also be second-order uncertainty, i.e. uncertainty concerning the extent of uncertainty.

Another important variable is the ambiguity of risk assessment (Renn 2008). Some risk assessments (vehicle accidents) are widely accepted by leaders and the population whereas others are not, even when based on scientific evidence (fish stock depletion, global warming). Risk can be acknowledged and action taken to mitigate it, but it can also be denied and business proceeds as usual. Even if acknowledged, risk can be reduced to a back-of-the-mind worry that does not result in safer practices, or it may be accepted in order to maintain present life-style or to access benefits. Although ambiguity is different from uncertainty, it typically arises under uncertainty and leads to debates concerning the acceptance of risk versus acting according to the precautionary principle, between being reckless or overly cautious.

These distinctions are ideal types helpful for the analysis that follows, but we should remember that in a specific empirical case there is usually a continuum between these poles as well as a complex combination. In this usage, 'risk society' refers to a society that creates new material dangers. This is conceptually different from, but empirically related to, what I would call the 'risk discourse society' where people increasingly worry and talk about risk and scientists make risk calculations. The second is an adjunct to the first. Paradoxically, the risk discourse society does not necessarily lead to a risk-minimising society. In many cases the contrary is true, with awareness of risk being relegated to a back-of-the-mind worry instead of an incitement to modify risky social practices. Visible, immediate, routine risks such as vehicle accidents, house fires and volcanic eruptions typically prompt protective or at least avoidance measures, such as better brakes on vehicles, smoke detectors in homes and evacuation from areas with active volcanoes (Murphy 2010). Difficult-to-perceive, slow-onset, uncertain risks can be denied, discounted or defined as acceptable even if they are high-impact risks, hence they are more challenging to manage.

Uncertainty, worst-case scenarios and resilience

Like other modern societies, Canada, Maine and northern New York state made themselves dependent on centralised electrical grids, roads, railways and airports. Until January 1998, meteorologists and power company authorities in north-eastern North America did not know the risk of five days of intense freezing rain over a vast area because it had never happened since records were kept (Murphy 2009). Such a storm was deemed highly unlikely, so the electrical grids in these areas were constructed to withstand only the lesser amounts of ice loading that occur regularly in winter. By becoming dependent on these infrastructures and by assuming safety based on short-term extrapolations, these societies inadvertently socially constructed vulnerability to this kind of force of nature. But the risk was present even if it was not anticipated. That unperceived risk was actualised on 5–9 January 1998. The ice loading from the five days of intense freezing rain crushed the electrical grids over a huge territory. The result was the most expensive disaster affecting the highest number of people in the history of each place. The population shivered without heat and light during the coldest part of winter, for almost a month for hundreds of thousands of people. Forty-seven people died, and modern production and way of life ceased. Not only harm but the extent of uncertainty were unforeseen – the main technical study of the disaster was entitled *Confronting the Unforeseeable* (CST 1999). Nature's construction of the worst case exceeded the social constructs of worst-case scenarios imagined by risk assessors. This demonstrates that the management of risk in conditions of uncertainty is incomplete unless we take into account the possibility that worst-case scenarios may be exceeded.

It has been argued that, in the face of uncertainty, the management of risk should rely upon resilience (Wynne 1992). Renn (2008, p. 179) contends that a principal objective of managing risks under uncertainty is to render systems resilient in the face of surprises: 'In contrast to robustness, where potential threats are known in advance and the absorbing system needs to be prepared to face these threats, resilience is a protective strategy against unknown or highly uncertain hazards'. The freezing rain event was a case laden with uncertainty, so what can be learned from it? When the risk assessments failed to anticipate the intensity, scope and duration of the extreme weather, the robustness built into the electrical grid to withstand extreme weather failed as well. The societies had to rely upon their resilience, i.e. on their capacity to bounce back after the electrical grids had been crushed. They proved to be resilient, but depending on resilience involved much human suffering and great expense in rebuilding what had been destroyed. The meaning of protection is deformed by calling resilience a protective strategy. It did not protect. Its only success was as a reconstruction strategy, but at great human and financial cost. This demonstrates the deficiencies of resilience as a way of coping with risks under uncertainty even when dealing with reversible phenomena. Furthermore, resilience functions only for wealthy, well-organised societies such as Canada and the US. Poor countries are not resilient, as was shown when an earthquake struck Haiti. Moreover, resilience defined as the capacity to bounce back to a previous state is an oxymoron when dealing with truly irreversible phenomena, which climatologists argue is the most worrisome threat of global warming (Schneider 2009).

Politics of climate change

There have been many environmental risks under conditions of uncertainty: ozone-layer depletion, biodiversity loss, acid rain, global warming etc. Some have been managed well and risks diminished, whereas others have not been and risks have been exacerbated. The current pre-eminent long-term environmental risk shrouded in uncertainties is arguably that of global climate change caused by human activities. Like seismologists for earthquakes, climatologists and other scientists have provided evidence of increased risk from global warming without being able to specify impact, location and timing of harm. Their knowledge is suggestive but partial, leaving room for it to be contested and ignored. Climatologists argue that modern industrial societies are running the risk of having their activities unleash climate dynamics that will push nature to a tipping point, where it will slide into irreversibility (Hansen *et al.* 2005; Schneider 2009; Flannery 2005). The paleontologist Flannery (2009) added that, even though visible consequences may not happen soon, the point of irreversibility could arrive quickly. Adaptation could be impossible if there are tipping points into a less benign environment. A parallel is the formerly popular morning-sickness drug thalidomide: consequences became visible only when deformed babies were born, but irreversibility for the fetus happened months earlier when the drug was taken. Adaptation was inferior to prevention for managing thalidomide problems. Those who seek to shift the emphasis from mitigation to adaptation (Lomborg 2001, 2007; Hulme 2009; Stehr and von Storch 2010) have to assume that climatologists are wrong. Hopefully climatologists will be proved wrong about irreversibility and urgency, but there is a significant risk that they are correct. The question then becomes whether uncertainty will be used as a reason for caution or for going full speed ahead.

The most renowned social theorist to contribute to debate concerning the management of climate change risks is the British sociologist Anthony Giddens. He contends that 'for most of the time and for the majority of citizens, climate change is a back-of-the-mind issue, even if it is a source of worry. It will stay that way unless its consequences become visible and immediate'(Giddens 2009, p. 106). There are two paths by which climate change can become a front-of-the-mind issue leading to action. The first is through the experience of disasters that clearly result from global warming. Disasters typically transform expectations about risk and usually (but not always) incite enhanced defences against similar hazards (Turner and Pidgeon 1978; Lewis 1999; Mileti 1999). Disasters would likely motivate mitigation of global warming, but surely there is a less painful way for social change to occur.

The second path is what Giddens calls 'foregrounding' climate change to make the politically non-feasible become feasible. How can foregrounding occur in a way that incites action to deal with risks like climate change under conditions of uncertainty? He claims that avoiding calamities constitute weak motivators. Opportunities, however, are strong motivators. This constitutes an indirect criticism of most environmental research since *Limits to Growth* (Meadows *et al.* 1972). Giddens (2009, p. 88) presents a specific example of how the pursuit of today's opportunities can decrease the risk of tomorrow's dangers: 'countries that are at the top of the league [of the climate change performance index] are there because of a preoccupation with energy security rather than climate change. In this sense, they have arrived where they are largely by accident'. Sweden, France, Denmark and Germany decreased their greenhouse gas (GHG) emissions because of concern over energy security rather than because they aimed to reduce global warming. The goal of energy security led states to promote more efficient use of fossil fuels and develop renewable energy and nuclear energy. The opportunities for jobs, profits and economic growth of these sectors indirectly motivated the reduction of GHG emissions. This made decreasing emissions more palatable to the population than sacrifices to diminish the risk of climate change. Giddens calls this political and economic convergence with the mitigation of climate change, and it is what he counsels to manage the risk of global warming with all its uncertainties.

Giddens' analysis is important. It recognises the high-impact consequences of the risks, the possibility of irreversible degradation of the presently beneficial natural environment and the high degree of uncertainty in assumptions of safety. Hence, it acknowledges that adaptation and resilience, although necessary, are inadequate and that mitigation is needed in cases like climate change that involve the possibility of irreversible tipping points. However, Giddens' analysis and suggestions for mitigation need to be seriously qualified. The politics of hope and an emphasis on opportunities have to be put on a solid footing in order to manage the risk that present practices are not sustainable in the long run. This is particularly challenging for slow-onset, high-impact, novel risks laden with uncertainty.

US energy security and greenhouse gas mitigation: convergence or divergence?

The US, once the greatest exporter of oil, is now the world's biggest importer. Its oil production peaked in 1970 and it now depends on imports for two-thirds of its oil (Hughes 2009). Politically stable, friendly Canada surpassed Saudi Arabia in 2004 as its main source of foreign oil. However, Canadian oil is mainly, and increasingly, extracted from tar sands. By 2009 'the oil sands mines have so far converted over 150 square miles of boreal forest … into dust, dirt, and tailings ponds' (Kunzig 2009, p. 54). Only a small fraction of the resulting brownfields have been restored. Extracting a barrel of crude oil from tar sands emits from 1.6 to three times more CO_2 than getting it from a conventional well (Kunzig 2009; Environment Canada 2011, p. 8). One barrel of energy equivalent is needed to produce five barrels of oil from tar sands deeper than 200 feet where 80% of the recoverable bitumen lies, whereas one barrel of energy yields 10 barrels of oil from conventional wells and once yielded 100 barrels from the Ghawar Field in Saudi Arabia (Hughes 2009). In Alberta, natural gas is used to supply the energy to extract oil from sand, which implies using relatively clean fossil fuel to produce dirtier fossil fuel. Tar sands exploitation uses huge quantities of fresh water, which become too contaminated to return to the river and are stored in toxic tailings lagoons. A recent study (Kelly *et al.* 2010) documented that the oil sands industry is contaminating the Athabasca River with elements that are toxic even at low concentrations. The biggest threat is that oil companies want to expand tar sands exploitation by a factor of three to five by 2020.

Despite the environmental degradation, the use of tar sands oil will contribute to US energy security. As the price of oil increases, consumption decreases somewhat, but dirty oil (from tar sands) and dangerous oil (from deepwater drilling and regions made accessible by global warming, such as the sensitive Arctic) become profitable and are increasingly used to attain energy security. The goals of energy security and mitigating GHG emissions thereby diverge, with the former contributing to global warming and risk unless dirty and dangerous sources of energy security are made clean and safe or are excluded. Giddens' solution needs to be qualified to avoid false hopes and wishful thinking.

Two different approaches compete to manage oil extraction from tar sands. The first involves the attenuation of risk. It consists of a substantive triage of technologies to reduce to conventional oil levels the absolute amounts of non-renewable energy used, fresh water used, clear-cutting of forests and GHG emissions. For example, a nuclear reactor instead of GHG-emitting natural gas could supply the energy for extracting the oil from the tar sands – if ever there were justification for the risks of a nuclear reactor, it is in the isolated area of the tar sands to reduce the risk of global warming. This approach would proceed in stages, requiring certification of the first stage before allowing a further stage of extraction and upgrading oil from tar sands. It would not imply the end of extracting oil from tar sands nor its contribution to US energy security, but would ensure it is done in an environmentally responsible way, with the requirements being incentives for the development of clean extraction and upgrading technology. This would reduce the contribution of tar sands oil extraction to the risk of global warming, but it has not been implemented because such a shift is novel and might be expensive.

The second approach involves limiting and framing communicative action to attenuate risk perception, and it is now under way. Tailings lagoons and clear-cuts are concealed from public view, easily done because the tar sands are far from the nearest city (Edmonton, 480 km away). Extraction areas are fenced off to reduce the visibility of environmental degradation, which can be seen only by helicopter. Allowing such pollution in northern Alberta is a forerunner to exploitation of the Arctic, which is even more isolated from population centres. Another element consists of a successful marketing campaign by the Canadian Association of Petroleum Producers to frame in a positive light the conversation on oil extraction from tar sands. Space limitations prevent description of its many aspects; two will have to suffice. The first is that emissions resulting from extracting and upgrading oil from sand are rhetorically minimised by shifting the focus to 'full cycle – well to wheels' level of emissions. Because emissions involved in transportation, refining and combustion of tar sands oil are identical to conventional oil, the 60–300% greater emissions of extraction and upgrading become reduced to 15–40% greater emissions in the full cycle (Kunzig 2009), which appears much lower even though the excess emissions going into the atmosphere are the same. The second is that, in international negotiations, the oil industry and the Alberta and Canadian governments have shifted the reference date from 1990 (specified in the Kyoto Protocol) to 2006, after a huge increase in oil sands emissions occurred.

Propaganda can, however, be turned against the propagandist. Al Gore noted that driving a Prius with tar sands oil produces the same life-cycle emissions as using conventional oil to drive a Hummer. If oil comes from tar sands, the much greater emissions at the beginning of the full cycle defeat measures to reduce emissions at the end. Stricter US fuel-efficiency regulations will not reduce emissions if its vehicle fleet becomes powered by fuel extracted from tar sands instead of conventional sources. Risk is determined by the social practices that produce it, in this case fossil fuel emissions. Whereas the first approach involves the attenuation of risk, the second consists of the attenuation of risk perception. It is crucial to keep in mind the difference between risk and its perception, between safety and assumptions of safety.

Methodologies to assess the management of risk under uncertainty

The preceding section demonstrates that Giddens' analysis is an oversimplification. The goal of energy security often diverges sharply from the goal of mitigating GHG emissions. The emphasis on opportunities and energy security is important, but for effective risk management it is necessary to assess whether they mitigate global warming without jumping to wishful conclusions.

How is the success of mitigation and the management of risk to be assessed? Different methods have been used. One indicator is discourse concerning mitigation and adaptation. For example, the present Conservative Canadian government claims that under its leadership Canada will be a world leader in restraining GHG emissions. But after five years in government, it has done little to combat Canada's GHG emissions. Mitigation and adaptation discourse, even plans and policies, can be very misleading indicators of successful risk management because they can be just lip service or shelved plans.

The second method is to base assessment on attempts to mitigate environmental problems. The Kyoto Protocol was an attempt at restraining GHG emissions, but world emissions are increasing. Relativist indicators show that there are now fewer emissions per barrel of oil produced from tar sands than before. However, emissions per barrel are still higher than for conventional oil, and overall emissions are increasing because more barrels of tar sands oil are being extracted. Attempts and relativist criteria give the illusion of improvement even as emissions increase. They constitute misleading ways to assess management of the risk of global warming, because the atmosphere is affected by the absolute level of emissions.

A third way is the only valid method – using absolute material indicators that the problem is being solved. For example, the Montreal Protocol restricting the use of chloro-fluoro carbons halted depletion of the ozone layer. Hence, that protocol can be evaluated as successful. On the other hand, the carbon content and temperature of the atmosphere keep rising, hence the Kyoto Protocol can validly be judged as insufficient. Only this third methodology provides a compelling way to assess whether the management of environmental risk is being successfully addressed. Giddens contended that the state has to act not only as facilitator and enabler of mitigation of global warming, but that 'it has to ensure that definite outcomes are achieved – most notably a progressive reduction in carbon emissions'(2009, p. 8). Such an 'ensuring state' requires absolute biophysical indicators that the problem is being solved, i.e. the third methodology.

Precautionary principle or cost–benefit analysis?

Giddens (2009) argues against the precautionary principle to manage catastrophic risk because it is hard to operationalise and is incoherent since risk-taking is needed for daily life and for innovation. He prefers cost–benefit analysis. However, cost–benefit analyses are also difficult to apply and are even more incoherent for risk management, for several reasons. First, present lifestyle predispositions and received values, powerful economic interests and wishful thinking typically lead risks and future costs to be underestimated and defined as acceptable, e.g. the BP oil spill in the Gulf of Mexico that occurred in May 2010. Second, estimates of costs and benefits are often based on insufficient evidence, especially for managing never-before-experienced risks such as climate change. Definitive proof of risk may only become available following a disaster (Turner and Pidgeon 1978). Cost-benefit projections are not rigorous under uncertainty, hence the available options are precaution or recklessness. Third, whereas benefits are immediate and financial (profits, jobs), environmental costs and harm are distant in time and space, and therefore discounted. For example, extracting oil from tar sands gives immediate profits and jobs but the costs of global warming are experienced in the future and in distant locations. Immediate financial benefits with distant dangers and costs promote the acceptance of risk rather than mitigate it. Freudenburg *et al.* (2009) documented how cost–benefit analyses used for decades led to constructions that increased vulnerability and risk in New Orleans, thereby exacerbating damage when Hurricane Katrina struck in 2005. Renn (2008) argued persuasively that the precautionary principle is needed to ensure safety where there is the possibility of harm that is irreversible, ubiquitous and pervasive in time. This is the case with climate change. Giddens (2009) did not deal adequately with such threats in his dismissal of the precautionary principle in favour of cost–benefit analysis. When uncertain but potentially irreversible dangers are being created by modern societies, as in the unleashing of climate change, the management of risk requires the precautionary principle even if it is difficult to apply.

Conclusion

The extreme weather event examined in this chapter (five days of freezing rain in northern US) demonstrated that when risk assessors assume they can calculate threats, including the worst-case scenario, nature's dynamics can exceed their expectations. The scenario that occurred was worse than the social construct that was imagined. This showed that there is a great deal of uncertainty in risk assessments, both lay and scientific, especially for significant features such as the impact, scope, duration and timing of hazards. Those mental models are attempts to approximate their material referent, but they are fallible and can underestimate risk. The study also showed that resilience, namely the capacity to bounce back after disaster, is not a protective strategy. At best it is a restorative strategy. Resilience has importance if and only if irreversibility is not involved, but even then it involves much suffering and huge expense, and is not effective for poor countries. The strategy of letting disaster happen then bouncing back, and letting global warming occur then adapting, is a risk-maximising strategy laden with uncertainties, especially for poor countries. Unfortunately, it seems to be most societies' *de facto* approach to climate change.

The underestimation of the risk of intense, persistent, widespread freezing rain by lay and professional assessors documented here shows the need to clearly distinguish mental models of risk from their material referent. This conclusion can also be drawn from the examination of the exploitation of the Canadian tar sands: practices, policies and marketing campaigns can aggravate risk while claiming to attenuate it.

Scientific risk assessments concerning global warming are uncertain in the sense of being suggestive rather than definitive and of being unable to predict impact, location and timing of threats. Such assessments of heightened risk suggest the possibility of irreversible tipping points, of ubiquitous danger and emergent threats that are long-lasting. Resilience and adaptation, although necessary, are likely to be inadequate and to involve much suffering. Giddens correctly acknowledged that mitigation of global warming is necessary and suggested a strategy to attain it indirectly, by aiming for energy security and opportunities. But that strategy is contingent. It has been successful in some contexts, such as in northern Europe, but is in the process of backfiring in North America. Risk awareness in the latter are being attenuated by economic and political interests even as the risk of global warming increases. It is necessary to anchor risk management in absolute biophysical indicators that clarify whether the problem is being solved or is worsening. This has important implications for social theory. Anchoring risk mitigation biophysically requires true interdisciplinary research that transcends the social science/natural science divide.

Acknowledgements

I would like to thank the Social Sciences and Humanities Research Council of Canada for a grant that supported this research, and the Environment and Society Research Committee of the Australian Sociological Association for funding to present a preliminary version of this chapter at its annual meetings.

References

Beck U (1992) *Risk Society.* Sage, London.

Beck U (2009) *World at Risk.* Polity Press, Cambridge.

Benedick R (1991) *Ozone Diplomacy: New Directions in Safeguarding the Planet.* Harvard University Press, Cambridge.

CST (Commission scientifique et technique chargée d'analyser les événements relatifs à la tempête de verglas) (1999) *Pour affronter l'imprévisible.* Gouvernement du Québec, Québec.

Diamond J (2005) *Collapse: How Societies Choose to Fail or Succeed.* Viking, New York.

Dray P and Cagin S (1993) *Between Earth and Sky: How CFCs Changed our World and Endangered the Ozone Layer.* Pantheon Books, New York.

Environment Canada (2011) 'National inventory report 1990–2009: greenhouse gas sources and sinks in Canada executive summary'. Environment Canada, Ottawa.

Flannery T (2005) *The Weather Makers: How We are Changing the Climate and What it Means for Life on Earth.* Text Publishing, Melbourne.

Flannery T (2009) *Now or Never.* HarperCollins, Toronto.

Freudenburg W, Gramling R, Laska S and Erikson K (2009) *Catastrophe in the Making: The Engineering of Katrina and the Disasters of Tomorrow.* Island Press, Washington.

Giddens A (2009) *The Politics of Climate Change.* Polity Press, Cambridge.

Gross M (2003) Sociologists of the unexpected. *American Sociologist* **34**, 40–58.

Hannigan J (1995) *Environmental Sociology.* Routledge, London.

Hannigan J (2006) *Environmental Sociology.* Routledge, London.

Hansen J, Nazarenko L, Ruedy R, Sato M, Willis J, Del Genio A, Koch D, Lacis A, Lo K, Menon S, Novakov T, Perlwitz J, Russell G, Schmidt GA and Tausnev N (2005) Earth's energy imbalance: confirmation and implications. *Science* **308,** 1341–1345.

Hughes D (2009) The energy issue: a more urgent problem than climate change? In *Carbon Shift: How the Twin Crises of Oil Depletion and Climate Change will Define the Future.* (Eds T Homer-Dixon and N Garrison) pp. 58–95. Random House, Toronto.

Hulme M (2009) *Why We Disagree about Climate Change: Understanding Controversy, Inaction and Opportunity.* Cambridge University Press, Cambridge.

Kelly E, Schindler D, Hodson P, Short J, Radmanovich R and Nielsen C (2010) Oil sands development contributes elements toxic at low concentrations to the Athabasca River and its tributaries. *Proceedings of the National Academy of Sciences* **107**, 16178–16183.

Kunzig R (2009) The Canadian oil boom: scraping bottom. *National Geographic* **215**, 34–59.

Lewis J (1999) *Development in Disaster-prone Places: Studies in Vulnerability.* Intermediate Technology Publications, London.

Litfin K (1994) *Ozone Discourses: Science and Politics in Global Environmental Cooperation.* Columbia University Press, New York.

Lomborg B (2001) *The Skeptical Environmentalist.* Cambridge University Press, Cambridge.
Lomborg B (2007) *Cool It.* Alfred A Knopf, New York.
Meadows DL, Meadows DH, Randers J and Behrens WW (1972) *The Limits to Growth.* New American Library, New York.
Mileti D (1999) *Disasters by Design.* Joseph Henry, Washington.
Murphy R (1999) Unperceived risk. *Advances in Human Ecology* **8**, 99–123.
Murphy R (2009) *Leadership in Disaster: Learning for a Future with Global Climate Change.* McGill-Queen's University Press, Montreal.
Murphy R (2010) Environmental hazards and human disasters. In *The International Handbook of Environmental Sociology.* (Eds M Redclift and G Woodgate) pp. 276–291. Edward Elgar, London.
Oliver-Smith A (2001) Theorizing disasters: nature, power, and culture. In *Catastrophe and Culture: The Anthropology of Disaster.* (Eds S Hoffman and A Oliver-Smith) pp. 23–47. School of American Research Press, Santa Fé.
Renn O (2008) *Risk Governance: Coping with Uncertainty in a Complex World.* Earthscan Publications, London.
Schneider S (2009) *Science as a Contact Sport: Inside the Battle to Save Earth's Climate.* National Geographic, Washington.
Short JF (1989) On defining, describing, and explaining elephants (and reactions to them): hazards, disasters, and risk analysis. *Mass Emergencies and Disasters* **7**, 397–418.
Simon J (1981) *The Ultimate Resource.* Princeton University Press, Princeton.
Stehr N and von Storch H (2010) *Climate and Society: Climate as Resource, Climate as Risk.* World Scientific Publishing, Singapore.
Turner B and Pidgeon N (1978) *Man-made Disasters.* Wykeham, London.
Vaughan D (1996) *The Challenger Launch Decision.* University of Chicago Press, Chicago.
Wynne B (1992) Uncertainty and environmental learning: reconceiving science and policy in the preventive paradigm. *Global Environmental Change* **2**, 111–127.

3 RISK AND ENVIRONMENTAL VICTIMISATION

Rob White

Introduction

A recent incident in Hungary provides a tragic illustration of what transnational environmental harm really means. A thick red torrent of toxic sludge burst from a reservoir at a metals plant 100 km south of Budapest in early October 2010. At least nine people died as a result of the sludge surge, some went missing and over 100 persons were physically injured as the toxic substance flowed into nearby villages and towns. The sludge reached the Danube River several days later, from where it could flow into six other European countries – Croatia, Serbia, Romania, Bulgaria, Ukraine and Moldova – before reaching the Black Sea. An ecological and social disaster for Hungary thus simultaneously posed an environmental threat to surrounding countries and their human inhabitants, ecosystems and animal life.

What happens in one country can, in many cases, affect people, ecosystems and animals well beyond that country's borders. Harm and risk of harm are frequently and increasingly transnational in character.

How to interpret, respond to and prevent such events is part of the mandate of social scientists with an interest in analysing existing and future threats to environmental well-being. Certainly matters of time, space and scale are relevant. For example, risks and harms may be direct or indirect, and their consequences may be felt in the immediate or in the long term. Harm may be specific to local areas (e.g. threats to certain species, such as coral in the Great Barrier Reef) yet manifest as part of a general global pattern (e.g. an effect of wide-scale temperature changes affecting coral everywhere). Harm is central, but this may be unintentional (in the sense of being a byproduct of some other agenda) or premeditated (insofar as the negative outcome, for some, is foreseen). The demise of polar bears due to the impact of global warming in the Arctic is an example of the former. The displacement of local inhabitants from their land due to carbon sequestration schemes is an example of the latter.

The intention of this chapter is to explore environmentally related risks by analysing current trends and patterns of victimisation including responses to victimisation, from the point of view of eco-global criminology (White 2011). Rather than being restricted by the limitations of the legal/illegal divide, this perspective asserts the precedence of 'the ecological' and ecological justice – the idea that ecological systems should be diverse and productive over time, and that there ought to be an equitable and just future for all. This means assessing 'harm' in many different contexts and guises, regardless of legal status and existing institutional legitimations. Questions of risk and potential harm are not simply technical in nature. Rather, as demonstrated in this chapter, risk tends to be apportioned to certain groups of people more than others.

Environmental risk and social inequality frequently go hand-in-hand. Moreover, substantive potential harms are increasing in likelihood as a result of global warming.

Environmentally related risks

For the green criminologist, the greatest threats to environmental rights, ecological justice and non-human animal well-being are system-level structures and pressures that commodify all aspects of social existence, based upon the exploitation of humans, non-human animals and natural resources, and that privilege the powerful minority over the interests of the vast majority (White 2008a). Those who determine and shape the law are very often those whose activities need to be criminalised for the sake of planetary well-being. Environmental harm is thus intrinsically contestable, both at the level of definition and in terms of visions of what is required for desired social and ecological change. Who says what, and whose voices are heard, become crucial in determining what is deemed to be harmful and which issues are most likely to garner attention.

Intervention in environmental matters partially depends upon how risk is conceived and whether assessment of risk subsequently leads to action. Responding to environmental harm is not only about reacting to specific events or incidents. It includes evaluation of potential threats or risks into the future. Taking precaution is central to protecting the planet, humans and non-human animals from projected harms. This involves weighing up and recognising which risks actually exist, and for whom.

Risk is a multi-dimensional entity generally incorporating several key elements. One notion of risk sees it as a prediction or expectation that involves a hazard (the source of danger), uncertainty of occurrence and outcomes (expressed by the probability or chance of occurrence), adverse consequences (the possible outcome), a time-frame for evaluation and the perspectives of those affected about what is important to them (Leiss and Hrudey 2005). Drawing upon a list provided by Deville and Harding (1997), a number of threats to the environment can be identified. This list is by no means exhaustive, but it does provide some indication of the types of threats that the precautionary principle aims to avoid:

- global warming (e.g. due to excessive discharges of carbon dioxide);
- biodiversity loss (e.g. due to release and establishment of non-native plant and animal species);
- stratospheric ozone depletion (e.g. due to use of CFCs);
- desertification and land degradation (e.g. due to land clearing for unsustainable agricultural practices);
- marine ecosystem health (e.g. due to oil spills);
- freshwater ecosystem health (e.g. due to discharge of pollutants);
- acid rain (e.g. due to atmospheric pollutants);
- damage to specific ecosystems (e.g. due to overfishing or overlogging);
- damage to human and non-human physical and mental health (e.g. due to chemical residues in food).

Close analysis and thoughtful reflection about this list generates a further set of questions that are relevant to issues surrounding environmental harm and victimisation. For example, who or what is the victim? What are the actual risks? What are the likely harms? Can precautions be taken in relation to these risks? What are the roles of scientific analysis and non-scientific understandings and experiences (e.g. layperson accounts) in assessing risks, and who, specifically, ought to take the requisite action? These and other questions deserve consideration in any discussion of risk and precaution. For present purposes the main concern is who bears the greatest risk of environmental harm.

Victimisation is central to the notions of 'risk' and 'precaution', since each is interpreted in terms of potential threat to human and environmental well-being. From the point of view of eco-global criminology, analysis of the nature of environmental harm has to take into account objective and subjective dimensions of victimisation. It also has to locate the processes of environmental victimisation within the context of the wider political economy. That is, the dynamics of environmental harm cannot be understood apart from consideration of who has the power to make decisions, the kinds of decisions that are made, in whose interests they are made and how social practices based on these decisions are materially organised. Issues of power and control also have to be analysed in light of global economic, social and political developments.

The 'choices' ingrained in environmental victimisation (of human beings, of the non-human world) stem from systemic imperatives to exploit the planetary environment for production of commodities for human use. This is not a politically neutral process. In other words, how human beings produce, consume and reproduce themselves is socially patterned in ways that are dominated by global corporate interests (Athanasiou 1996; White 2002).

The dominance of neo-liberal ideology as a guiding rationale for further commodification of nature and the concentration of decision-making in state bureaucracies and transnational corporate hands accelerate the rate and extent of environmental victimisation. The power of capitalist hegemony manifests in the way in which certain forms of production and consumption become part of a taken-for-granted common sense, the experiences and habits of everyday life (see White 2010, regarding capitalism and the transformation of nature).

What gets defined as 'harm' or as a 'risk' is contingent upon the capacity of sectional interests to secure measures for generalising and implementing action against what is deemed to be 'harmful' behaviour, primarily via the state. Harm is rendered invisible to the extent that it can be externalised to more vulnerable population groups that do not have the social and political networks of influence equivalent to those of the powerful. These issues are worthy of close attention.

Patterns of victimisation

Environmental victimisation is not a socially neutral process, in that some people suffer more than others when it comes to poor environmental living conditions and/or events that are disastrous to their lives.

> Many have noted that there is a direct relationship between the increasing globalisation of the economy and environmental degradation of habitats and the living spaces for many of the world's peoples. In many places where Black, minority, poor or Indigenous peoples live, oil, timber and minerals are extracted in such a way as to devastate ecosystems and destroy their culture and livelihood. Waste from both high- and low-tech industries, much of it toxic, has polluted groundwater, soil and the atmosphere. The globalization of the chemical industry is increasing the levels of persistent organic pollutants, such as dioxin, in the environment. Further, the mobility of corporations has made it possible for them to seek the greatest profit, the least government and environmental regulations, and the best tax incentives, anywhere in the world (Robinson 2000, quoted in White 2008a, p. 116).

The notion of environmental victim implies that someone or something is being harmed through the conscious or neglectful actions of another. From an eco-global criminology perspective, environmental harm is best seen in terms of justice, based upon notions of human, ecological and animal rights, and egalitarian concerns. The key endeavour is the weighing up of different forms of harm and violation of rights within a broad eco-justice framework, and stretching the boundaries of conventional criminology to include other kinds of harm than those already deemed illegal. Environmental victimology as such, however, is less concerned with non-human animals and specific biospheres than with the interests and well-being of humans in specific circumstances.

According to Williams (1996, p. 21), the term 'environmental victim' represents the idea of injury caused by a deliberate or reckless act or omission. Environmental victims are 'those of past, present, or future generations who are injured as a consequence of change to the chemical, physical, microbiological, or psychosocial environment, brought about by deliberate or reckless, individual or collective, human act or act of omission'. Environmental victimisation can be defined as specific forms of harm that are caused by acts or omissions leading to the presence or absence of environmental agents which are associated with human injury (Williams 1996). It is important to distinguish between environmental issues that affect everyone and those that disproportionately affect specific individuals and groups (Williams 1996; Low and Gleeson 1998). In some instances, there may be a basic equality of victims in that some environmental problems threaten everyone in the same way, for example in the case of ozone depletion, global warming, air pollution and acid rain (Beck 1996).

However, as extensive work on specific incidents and patterns of victimisation demonstrates, it is also true that some people are more likely to be disadvantaged by environmental problems than others. For instance, studies have identified disparities involving many different types of environmental hazard that especially adversely affect people of colour, ethnic minority groups and indigenous people in places such as Canada, Australia and the US (Bullard 1994; Pinderhughes 1996; Langton 1998; Brook 2000; Rush 2002). There are patterns of differential victimisation evident with respect to the siting of toxic waste dumps, extreme air pollution, chemical accidents, access to safe clean drinking water and so on (Chunn *et al.* 2002; Saha and Mohai 2005; Williams 1996). It is the poor and disadvantaged who suffer disproportionately from such environmental inequalities.

There are also differential risks within at-risk populations. This is particularly apparent when it comes to socially differentiated characteristics such as age (the very young, the very old), gender (women tend to suffer more than men in circumstances of famine and disaster) and general health (mental, physical, ability/disability). For instance, a particular suburb or city may be placed in circumstances that heighten risks to well-being and health for everyone (e.g. dumping of toxic waste in Abidjan in Ivory Coast, the spraying of chemical pesticides in New York City). However, where heightened risk is deemed to be 'acceptable', particularly in terms of cost–benefit analysis (e.g. the use of pesticides to prevent the spread of disease borne by mosquitoes) there are hidden costs that may not be factored in. For instance, children and those with chemical sensitivities will suffer disproportionately if chemicals are sprayed since they are more vulnerable than others to ill effects arising from these toxins. In such circumstances, the crucial questions are not only 'How many will be harmed' but 'Who will be harmed'? (Scott 2005, p. 56). To appreciate this, we need to be conscious of differences (as well as similarities and commonalities) within affected populations.

Who is a victim includes differing degrees of harm, injury and suffering. Death from environmental catastrophe is only one manifestation of victimisation. Whether the affliction is or incorporates a disease or permanent injury or prolonged mental illness and psychological distress, a large proportion of 'victims' are simultaneously 'survivors'. They sometimes sustain injuries that significantly alter the course and quality of their lives and that are economically onerous in terms of health care. The breaking up of communities, the displacement of individuals, the loss of economic livelihood and dispossession of land all constitute varying forms and degrees of harm to human populations.

As indicated above, one of the tenets of victimology is that being and becoming a victim is never socially neutral. This holds true for environmental victimisation as it does for other sorts of victim-making. Fattah (2010, p. 46) notes that:

> In most instances victims are not chosen at random, and in many cases the motives for the criminal act develop around a specific and nonexchangeable victim. Therefore, an examination of victim characteristics, of the place the victim occupies, and the role the victim plays in these dynamic processes is essential to understanding why the crime was committed in a given situation, at a given moment, and why a particular target was chosen.

This should not be interpreted as suggesting that the victim is in some way responsible for their targeting. Rather, it is an acknowledgement that the more we examine specific actions that produce and involve environmental victims, the more it seems that those who suffer harm do so because of their specific relationship to the perpetrators of the harm. Largely these consist of relations of power, domination and exploitation. It is the social, economic and political characteristics of victim populations that make them vulnerable to victimisation in the first place.

Accordingly, environmental justice discourse places inequality, and inequalities in the distribution of environmental quality, at the top of the environmental agenda (Julian 2004; Harvey 1996). After all, environmental victimisation generally involves, on the one hand, powerful players such as corporations and nation-states and, on the other, less powerful groups such as indigenous people, ethnic minorities, the poor and those less able to take care of their own interests (e.g. the elderly and the very young).

Consider the environmental impact of mining on the Ok Tedi River in Papua New Guinea. Over a period of many years, the BHP mine operators discharged 80 million tonnes of tailings, overburden and mine-induced erosion into the river system each year. This caused widespread harm to the 50 000 people who live in the 120 villages downstream from the mine. The main beneficiaries from the mine have been BHP, the PNG government and the clan that holds ownership of the land on which the mine is built. Experts have predicted that it will take 300 years to clean up the toxic contamination (Kirsch 2006; Low and Gleeson 1998).

Another illustrative example is persistent pollution in the surrounding areas of a residential community. In the historic African American community of Mossville, Louisiana, it has been documented that residents have an average level of dioxins three times higher than that of the general US population. Dioxins are the most toxic chemicals known to science and can cause cancer, reproductive damage and extensive harm to foetal and child development. Independent analysis of government data has demonstrated that six local industrial facilities are the sources of the elevated dioxin levels in the Mossville community. The facilities release millions of tonnes of toxins into the air, water and land each year. Yet government agencies continue to issue permits which allow the industrial facilities to increase the amount of toxic pollution, including dioxins, which they release into the local community (Mossville Environmental Action Now 2007).

Such instances of blatant disregard for human health and well-being have been labelled environmental racism. To those who engage in systematic study of such questions, it is clear that, regardless of intent, the practical outcome of corporate and government action has been that disadvantaged groups end up living in the most hazardous and environmentally poor areas (Pellow 2007). This is so whether it occurs in the US (Bullard 1994), Canada (Chunn *et al.* 2002), India (Engel and Martin 2006) or Australia (Walker 2006). These kinds of communities also suffer most from the extraction of natural resources. Specifically, in many places around the globe where minority or indigenous peoples live, oil, timber and minerals are extracted in ways that devastate local ecosystems and destroy traditional cultures and livelihoods (Schlosberg 2004, 2007).

Global warming and environmental victimisation

The context of global warming, declining oil resources and food crises puts even more of the world's ecological and economic burdens on the backs of the poor. As Shiva (2008, pp. 5–6) observed:

> First, they are displaced from work; then they bear a disproportionate burden of the costs of climate chaos through extreme droughts, floods, and cyclones; and then they lose once more when pseudo-solutions like industrial biofuels divert their land and their food. Whether it is industrial agriculture or industrial biofuels, car factories or superhighways, displacement and forced evictions of indigenous peoples and peasants from the land are an inevitable consequence of an economic model that creates growth by extinguishing people's rights.

Displacement from homelands is accelerating through the acquisition by foreign governments and private companies of large areas of arable land in developing countries.

> Parcels of several hundred thousands of hectares are being bought or leased in Africa, Central and Southeast Asia, and Eastern Europe by food-importing countries with domestic land and water constraints but abundant capital, such as the Gulf States, and by countries with large populations and food security concerns, such as China, South Korea and India (Sutherland *et al.* 2009, p. 5).

These land acquisitions are having major negative impacts on local people who are losing access and control over the resources on which they depend and which are the rightful inheritance of future generations.

These problems are compounded in some cases by the ways in which 'conservation' is being foisted upon these same communities. As Duffy (2010, p. 11) pointed out:

> When wildlife reserves are established, local communities can suddenly find that their everyday subsistence activities have been outlawed and they have been redefined as criminals ... Some of the world's best-known pristine wilderness areas are, in fact, engineered environments. Creating a national park means drawing up new conservation rules which outlaw the everyday subsistence activities of local communities, such as hunting for food and collecting wood.

In this way, victims can be transformed into offenders. A further example is climate-induced subsistence activities that do further harm to the environment. The plight of the dispossessed and disadvantaged means that often any environmental destruction brought about by their actions (cutting down of forests, fishing in protected areas or another state's exclusive economic zone) is best remedied by social justice initiatives rather than by criminal justice interventions. A person with no land and no natural resources already faces a huge and daunting task to survive – being subjected to ill treatment and placed in prison (i.e. detention centres) constitutes an additional harm that violates their very being.

Certain groups of people have histories of victimisation relating to the imposition of colonial power, which was intrinsically a matter of resource colonisation, a phenomenon that affected many different indigenous peoples in places such as South America, North America and Australasia as well as the native inhabitants of Africa, Asia and beyond. In places such as Australia, indigenous territories were considered frontier lands that were un-owned, under-utilised and therefore open to exploitation. The prior ownership rights, interests and knowledge of indigenous inhabitants were treated as irrelevant by the European invaders. The disregard for the physical and cultural well-being of indigenous people and their connection to 'country' was also evident in how they and their lands were treated when it came to nuclear testing and in establishing mining interests (White and Habibis 2005). Environmental victimisation has been central to dispossession and maltreatment of indigenous peoples across many continents and several centuries.

Such processes continue today, albeit under new guises. For example, one consequence of climate change has been the policy search in the west for band-aid solutions to carbon emissions. There are two kinds of disparities associated with this. First, most emissions of the North are 'luxury' emissions and contrast sharply with the 'survival' emissions of the South. Second, although not the main perpetrators of the problem, the people of the South suffer most from the effects of disasters related to climate change (see below).

The issue of carbon colonialism is also emerging as a form of climate injustice. This relates to the harnessing of forests as carbon sinks (by commercial interests in developed countries) for which credits can be earned and finance provided to developing countries, particularly biodiverse countries such as those in Latin America. The governments of such countries stand to gain financially from such arrangements, often at the expense of local communities. It has been alleged, for example, that a Norwegian company operating in Uganda leased its lands for a sequestration project that is said to have resulted in 8000 people in 13 villages being evicted (Bulkeley and Newell 2010, p. 48). Frances Seymour from the Center for International Forestry Research claimed that:

> as payments for conserving forests for carbon storage become increasingly likely, state and non-state actors alike will have strong incentives to passively ignore or actively deny the land and resource rights of indigenous, traditional and/or poor forest users in order to position themselves to claim compensation for forest stewardship in their stead (quoted in Bulkeley and Newell 2010, p. 77).

The effects of the commodification of carbon (i.e. its extraction from local contexts and circulation in global markets) will thus put pressure on communities that are home to these resources.

Those who are most vulnerable to the effects of climate change are the same groups who are vulnerable to events such as droughts, floods and cyclones. The conventional approach to disasters is to see them as 'natural' (and to include such things as earthquakes, volcanoes and floods) or human-caused (relating to fires, explosions and oil spills) (Picou *et al.* 2009). In the context of major global changes in climate, biodiversity and pollution, this presumption may no longer be warranted.

Similarly, Williams (1996) argued that when formulating a definition of environmental victim it is usually necessary to exclude those who are more accurately defined as 'environmental casualties', as this relates to those who have suffered as a result of natural disasters. This suffering, it is argued, is the result of chance. However:

> some circumstances that appear natural may, if analyzed in greater depth, be a consequence of human acts … Environmental suffering that has affected many generations, such as iodine deficiency, might not be seen as victimization until power relationships are examined. Why are the communities that suffer iodine deficiency forced to live on land that cannot sustain human life properly? (Williams 1996, p. 19).

This same logic extends to the causes and impact of specific events. Again, those least responsible for climate change are worst affected by it.

> Peasants, indigenous peoples, and artisans who live outside the industrialized globalized economy, who have caused no harm to the earth or other people, are the worst victims of climate chaos. Over 96% of disaster-related deaths in recent years have taken place in developing countries. In 2001, there were 170 million people affected by disasters around the world, of which 97% were climate-related (Shiva 2008, p. 3).

These vulnerabilities to victimisation are due not only to geographical location but to other inequalities. For example, many countries have coastal areas that are vulnerable to sea-level rise. But the Netherlands has the technological and financial capacity to protect itself to a greater extent than does Bangladesh. Thus, not only are poorer countries less responsible for the problem, they are simultaneously least able to adapt to the climate impacts they will suffer because they lack the resources and capacity to do so. This raises three key questions surrounding matters of justice:

- the question of responsibility (e.g. the North owes the South an 'ecological debt');
- the question of who pays for action on mitigation and adaptation;
- the question of who bears the costs of climate governance (Bulkeley and Newell 2010).

From the point of view of environmental activism and advocacy, the fact that environmental victims are frequently drawn from the ranks of the poor, the disadvantaged and the minority has significant ramifications. For example, many such victims fit into the category of 'socially expendable victims' (Fattah 2010). That is, no one really cares what happens to these specific individuals and groups, since they are already devalued in wider community terms. As Engel and Martin (2006, p. 479) put it: 'If victims are perceived as degraded in some sense, then it does not seem so unfair when bad things happen to them.'

This is in addition to the existing stigmatisation of 'ordinary' victims. The term 'victim' is a word that 'evokes strong images of submissiveness, pain, loss of control and defeat … Victims are riddled with taboos' (Rock 2007, p. 41). To be judged on preconceptions based upon race or income adds further fuel to the debasement associated with victimisation processes. To (re)act against perpetrators of environmental harm (which typically include powerful forces and organisations) often means first throwing off the chains of injustice and underprivilege that, in turn, undermine confidence and tactical wherewithal.

Differential victimisation relates to the subjective disposition and consciousness of the people involved in other ways as well. The specific groups who experience harm may not always describe or see the issues in strictly environmental terms. This may be related to lack of awareness of the environmental harm, alternative explanations for the calamity (e.g. an act of God) and socio-economic pressures to accept environmental risk in return for economic reward (Julian 2004). Waldman (2007), for instance, described a local community in South Africa that saw the contamination effects of asbestos as natural. This was due to a combination of religious beliefs (that stressed a passive stance to the world around them) and the fact that harms that are imperceptible to the senses often exist as a problem only if they are constituted as such in public discourse (in particular, the public discourse of the village community). Otherwise, what is, simply is.

This last point is important in another respect. Those who study disasters, for instance, often make a distinction between human-made (or technological) disasters and natural disasters (Picou *et al.* 2009). The latter are seen as acts of God and are responded to very differently from the former. Something that is seen to be naturally caused or created tends not to generate the same anger, angst and conflict as that which is perceived to be due to human error and/or conscious intervention. On the other hand, human-caused disasters frequently end up in drawn-out resolution processes that, in their own way, stifle or diminish victim resolve over time.

Nevertheless, environmental justice movements have historically tended to stem from specific instances of environmental victimisation (Bullard 1994). Such movements largely focus on redressing the unequal distribution of environmental disadvantage. Action may also be taken to prevent environmental hazards being located in particular local areas. For example, in the 1990s the communities of the lower Fly Region in Papua New Guinea sued BHP and received an out-of-court settlement, which was the culmination of an enormous public-relations campaign against the company by environmental groups. In January 2007, another lawsuit was lodged on behalf of other villagers seeking billions of dollars in damages.

There are different types of environmental victimisation (for elaboration and examples see Stevens 1996; White 2009, 2010). Environmental victimisation may be direct or indirect, immediate or long-lasting, local or regional. It may involve lead in soils, dioxins in water, radioactivity in the atmosphere. It may be based upon routine industrial practices or specific events such as climate-related disasters. The threat may be realised (due to actual presence or absence of something in the environment) or potential (e.g. a proposed privatisation of drinking water, development plans to build a dam or pulp mill). Children are much more vulnerable to some types of environmental harm (e.g. toxic chemicals) than are adults. In some cases, victimisation is more a question of proximity to the harm (e.g. death and maiming related to explosions, poisoning related to industrial emissions).

Responding to environmental risk and victimisation

At the heart of investigations of transnational environmental victimisation is the question of whose knowledge of 'wrong' is right? In other words, whose voices are going to be heard, and to what kinds of evidence do we lend credibility? It is rare that scientific evidence is uncontested or that proof of environmental harm is simply a matter of 'let the facts decide'. What counts as 'science', what counts as 'evidence', who counts as being a 'scientific expert' and what counts as 'sensible' public policy are all influenced by factors such as economic situation, the scientific tradition within a particular national context, the scientific standards used in relation to specific issues, and the style and mode of government (White 2008a, b).

In responding to environmental victimisation there are inevitably a range of social interests and discourses that contribute to the shaping of perceptions and issues (Hannigan 2006). This implies differences in perspective and a certain contentiousness of knowledge about the nature of the harm or crime. Assessment of victimisation usually involves responding to a series of interrelated questions.

- How are 'harm' and 'risk' defined, and by whom?
- How do we distinguish 'risk' (potentials) from 'harm' (actuals)?

- At what point does 'risk' or 'harm' actually occur?
- At what point does 'risk' or 'harm' occur to an extent warranting action or intervention?
- What is the origin of the harm?
- What are the histories of 'risk' or 'harm'?
- Why and how did the harm occur?
- Who are the perpetrators and why did they do it?
- Is responsibility in the hands of those harming, or of those being harmed?
- Who has responsibility for proof?
- What is acceptable as evidence?
- What were the triggers for the harm?
- What are the immediate signs of 'danger' or 'harm'?
- What strategies were invoked to diminish or mitigate the harm?
- How do we stop the harm from occurring again?

Answers to these questions frequently vary, depending upon stakeholder perspective. Consider, for example, the variety of players who might be associated with disputes over toxic landfill in a residential community adjacent to a mining operation. Because victimisation is a contestable social process that engages a wide range of individuals it is important to identify stakeholders and their interests (e.g. workers and jobs, residents and amenity). It is also useful to explore the diverse and often conflicting discourses around risk and harm from different stakeholders (e.g. medical risk vis-à-vis the health department, loss of livelihood in the case of farmers, limited perception among local miners that there is a problem).

The marshalling of particular types of evidence is typically driven by very specific requirements (and forms of evidence) related to institutions and groups. Who says what and why is linked to specific social purposes and particular discursive domains. The language of crime and victimisation is reflective of how an environmental problem (in this case toxic landfill) is socially constructed depending upon how it is being considered and by whom.

Just as environmental victimisation differs in manifestation, so do victim responses vary greatly. In broad terms, different events in different countries give rise to responses that vary from the passive to the confrontational, and from those involving collaborative activities aimed at redress to those based upon violence (Williams 1996).

Detailed analysis of specific events, over time, reveals stages in victims' struggle for justice, involving both spontaneous and organised actions, usually centred upon justice and/or relief. The effectiveness of specific struggles, such as those related to events in Bhopal, India or the Cape region in South Africa (Waldman 2007) can be analysed in terms of who defines the issues, who fights for or against the issues, who owns the struggle and how the struggle is shaped and carried out by local and international participants (Sarangi 1996; Waldman 2007; Engel and Martin 2006). Examination of victim responses needs to take into account the type and extent of networking and coalition-building, as well as the lack of participation and the marginalisation of some victim groups within a wider victim movement (Waldman 2007).

A related issue is how NGOs from 'outside' (the metropoles of the North) affect the status and livelihoods of those in certain parts of the world (those who live in the South). Duffy (2010) recounted how, in a number of cases, transnational NGO action translated into the criminalisation of local residents and their alienation from their lands and natural resources. There is a dark side to conservation, based upon cultural ignorance, that can create more harm than good and lead to both human misery and unsustainable ecological solutions.

Conclusion

This chapter takes the view that environmental victimisation is not a socially neutral process, but is constructed in the intersection between social inequality, unequal power relationships and the subjective experience of the victim. The majority of victims of environmental degradation – stemming from industrial and commercial activities, global warming, loss of biodiversity and increased waste and pollution – are very often the poor and the dispossessed. While all are threatened by global environmental disaster, there remain large social differences in the likelihood of and resilience to injury, harm and suffering. For those who disproportionately bear the brunt of global patterns of environmental transformation, degradation and victimisation, there are big questions on who will compensate them for their often prolonged suffering, now and into the future.

In both North and South, environmental victimisation could likewise occur but social interests and material comforts can provide protection for the privileged few, thereby precluding the thought of 'being a victim'. For the time being, victimisation will be experienced by these people as a matter of unlucky chance – until the day when chance is all that defines the moment. Until then, risk of environmental victimisation is still a relatively distant concern.

Importantly, the logic of risk assessment itself may serve to undermine community engagement on critical environmental issues and their consequences, particularly for those beyond the local landscape. This is because the framing of such issues mainly or solely in terms of risk implies that key questions relate to ascertaining the acceptable level of risk and determining what controls can be imposed to keep that risk within defined economic or ecological limits. Field (1998, p. 76) argued, for instance, that 'there is the danger that the debate will become mired in a highly technical discourse over the extent of risk and will lose sight of the equally profound issue of democratic control over the economic aspects of community life which is also presented by this [the environmental justice] movement'.

In other words, to assume risk is to ignore who produces risk. The danger lies in focusing on risk to the exclusion of those who produce risk and the links to the nature of production itself. The problem is not one of management and control, but of basic decisions pertaining to the means of life (and threats to this). Dealing with environmentally related risks demands action that goes to the source of the issues and the systemic imperatives that drive environmental degradation and global warming.

When sustained consideration is given to the characteristics and social processes of environmental victims, the patterns of victimisation that emerge demand a justice-based approach to questions of environmental risk. This, in turn, challenges risk logic, and the powerful social institutions and global structures that sustain it. Such considerations are at the heart of an eco-global criminology approach to harm.

References

Athanasiou A (1996) *Divided Planet: The Ecology of Rich and Poor.* Little Brown & Co., Boston.

Beck U (1996) World risk society as cosmopolitan society? Ecological questions in a framework of manufactured uncertainties. *Theory, Culture, Society* **13**, 1–32.

Brook D (2000) Environmental genocide: native Americans and toxic waste. *American Journal of Economics and Sociology* **57**, 105–113.

Bulkeley H and Newell P (2010) *Governing Climate Change.* Routledge, London.

Bullard R (1994) *Unequal Protection: Environmental Justice and Communities of Color.* Sierra Club Books, San Francisco.

Chunn D, Boyd S and Menzies R (2002) 'We all live in Bhopal': criminology discovers environmental crime. In *Toxic Criminology: Environment, Law and the State in Canada.* (Eds S Boyd, D Chunn and R Menzies) pp. 7–24. Fernwood Publishing, Halifax.

Deville A and Harding R (1997) *Applying the Precautionary Principle.* Federation Press, Sydney.

Duffy R (2010) *Nature Crime: How We're Getting Conservation Wrong.* Yale University Press, New Haven.

Engel S and Martin B (2006) Union Carbide and James Hardie: lessons in politics and power. *Global Society* **20**, 475–490.

Fattah E (2010) The evolution of a young, promising discipline: sixty years of victimology, a retrospective and prospective look. In *International Handbook of Victimology* (Eds S Shoham, P Knepper and M Kett) pp. 43–94. CRC Press, Boca Raton.

Field R (1998) Risk and justice: capitalist production and the environment. In *The Struggle for Ecological Democracy: Environmental Justice Movements in the US.* (Ed D Faber) pp. 69–94. Guilford Press, New York.

Hannigan J (2006) *Environmental Sociology.* 2nd edn. Routledge, London.

Harvey D (1996) *Justice, Nature and the Geography of Difference.* Blackwell, Oxford.

Julian R (2004) Inequality, social differences and environmental resources. In *Controversies in Environmental Sociology* (Ed R White) pp. 113–131. Cambridge University Press, Melbourne.

Kirsch S (2006) *Reverse Anthropology: Indigenous Analysis of Social and Environmental Relations in New Guinea.* Stanford University Press, Stanford.

Langton M (1998) *Burning Questions: Emerging Environmental Issues for Indigenous Peoples in Northern Australia.* Centre for Indigenous Natural and Cultural Resource Management, Darwin.

Leiss W and Hrudey S (2005) On proof and probability: introduction to 'law and risk'. In *Law and Risk* (Ed Law Commission of Canada) pp. 1–19. UBC Press, Vancouver.

Low N and Gleeson B (1998) *Justice, Society and Nature: An Exploration of Political Ecology.* Routledge, London.

Mossville Environmental Action Now (2007) *Industrial Sources of Dioxin Poisoning in Mossville, Louisiana: A Report Based on the Government's Own Data.* (Eds Wilma Subra and Advocates for Environmental Human Rights). Environmental Action Now, Mossville.

Pellow D (2007) *Resisting Global Toxics: Transnational Movements for Environmental Justice.* MIT Press, Cambridge.

Picou J, Formichella C, Marshall B and Arata C (2009) Community impacts of the Exxon Valdez oil spill: a synthesis and elaboration of social science research. In *Synthesis: Three Decades of Research on Socioeconomic Effects related to Offshore Petroleum Development in Coastal Alaska* (Eds S Braund and J Kruse) pp. 279–310. US Dept of the Interior, Anchorage.

Pinderhughes R (1996) The impact of race on environmental quality: an empirical and theoretical discussion. *Sociological Perspectives* **39**, 231–248.

Rock P (2007) Theoretical perspectives on victimisation. In *Handbook of Victims and Victimology* (Ed S Walklate) pp. 37–61. Willan Publishing, Devon.

Rush S (2002) Aboriginal resistance to the abuse of their national resources: the struggles for trees and water. In *Toxic Criminology: Environment, Law and the State in Canada.* (Eds S Boyd, D Chunn and R Menzies) pp. 67–86. Fernwood Publishing, Halifax.

Saha R and Mohai P (2005) Historical context and hazardous waste facility siting: understanding temporal patterns in Michigan. *Social Problems* **52**, 618–648.
Sarangi S (1996) The movement in Bhopal and its lessons. *Social Justice* **23**, 100–108.
Schlosberg D (2004) Reconceiving environmental justice: global movements and political theories. *Environmental Politics* **13**, 517–540.
Schlosberg D (2007) *Defining Environmental Justice: Theories, Movements, and Nature.* Oxford University Press, Oxford.
Scott D (2005) When precaution points two ways: confronting 'West Nile Fever'. *Canadian Journal of Law and Society* **20**, 27–65.
Shiva V (2008) *Soil not Oil: Environmental Justice in an Age of Climate Crisis.* South End Press, Brooklyn.
Stevens S (1996) Reflections on environmental justice: children as victims and actors. *Social Justice* **23**, 62–86.
Sutherland WJ, Clout M, Côté IM, Daszak P, Depledge MH, Fellman L, Fleishman E, Garthwaite R, Gibbons DW, de Lurio J, Impey AJ, Lickorish F, Lindenmayer D, Madgwick J, Margerison C, Maynard T, Peck LS, Pretty J, Prior S, Redford KH, Scharlemann JPW, Spalding M and Watkinson AR (2009) A horizon scan of global conservation issues for 2010. *Trends in Ecology and Evolution* **25**, 1–7.
Waldman L (2007) When social movements bypass the poor: asbestos pollution, international litigation and Griqua cultural identity. *Journal of Southern African Studies* **33**, 577–600.
Walker C (2006) Environmental racism in Australia. Chain Reaction No. 96, Autumn 2006. Friends of the Earth Australia magazine. <http://www.foe.org.au/mainfiles/cr.htm> (accessed 10.07.10).
White R (2002) Environmental harm and the political economy of consumption. *Social Justice* **29**, 82–102.
White R (2008a) *Crimes Against Nature: Environmental Criminology and Ecological Justice.* Willan Publishing, Devon.
White R (2008b) Depleted uranium, state crime and the politics of knowing. *Theoretical Criminology* **12**, 31–54.
White R (Ed) (2009) *Environmental Crime: A Reader.* Willan Publishing, Devon.
White R (2010) Environmental victims and resistance to state crime through transnational activism. *Social Justice* **36**, 46–60.
White R (2011) *Transnational Environmental Crime: Toward an Eco-Global Criminology.* Willan/Routledge, London.
White R and Habibis D (2005) *Crime and Society.* Oxford University Press, Melbourne.
Williams C (1996) An environmental victimology. *Social Justice* **23**, 16–40.

4 THE LURE OF THE MARKET IN TACKLING GLOBAL WARMING

Fiona Haines

> Labor[1] believes that to reduce the carbon pollution that causes climate change Australia must implement the Carbon Pollution Reduction Scheme. The CPRS delivered by Labor in government will for the first time put a price on carbon to encourage major polluting businesses to lower their emissions and invest in new, clean, low-pollution technologies (*Australian Labor, National Platform and Constitution 2009*).

> The CPRS is increasingly looking like the answer to a question that nobody asked, namely, what would be the best way to introduce a complex and expensive national scheme that sounds like a solution to climate change without really changing anything? (*Dr Richard Dennis, Executive Director of the Australia Institute, Wednesday 21 October 2009*).

> Labor's approach to climate change is nothing more than a great big tax on families and the Australian economy. It is a tax policy masquerading as an environment policy (*Coalition's[2] Direct Action Plan, July 2010*).

Introduction

Market mechanisms to stimulate a low carbon economy feature prominently in international efforts to reduce global warming. One such mechanism, emissions trading, has a particularly high profile with the major international trading regime set up under the Kyoto Protocol commencing trading in 2005. The largest regional market, the European Union Emissions Trading System (EU-ETS) began in the same year (Hepburn 2007). A proposal for an Australian trading regime, named the Climate Pollution Reduction Scheme (CPRS), was foreshadowed as the most significant initiative in the reduction of that nation's emissions in a major review into climate change led by Professor Ross Garnaut, a prominent Australian economist. The Garnaut Climate Change Review (hereafter Garnaut Review) envisaged a carbon emissions trading scheme (ETS) as the 'centrepiece of a domestic mitigation strategy' (Garnaut 2008b, p. 5). This policy, it argued, would bring Australia into line with the UN and EU emissions trading initiatives. Such endorsement was persuasive. Indeed, even before the Garnaut Review was finally published, the 2007 Rudd Labor government made the primary plank of its climate policy a national carbon emissions trading regime. Yet, despite this strong start, the proposal for an ETS in Australia failed to gain the necessary political and parliamentary support.

This chapter argues that the failure of the CPRS and the policies that replaced it reveal critical risk concerns, in particular the challenge of overcoming political risk in effective responses to anthropogenic climate change. The fortunes of emissions trading regimes and in particular the CPRS, from its inception to defeat, are traced in the quotes above and expanded on below. Following this defeat, there was a flurry of activity from each of the major political parties – Australian Labor, the Liberal/National Coalition and the Australian Greens. In the shadow of the 2010 Australian federal election, each party vied for public attention with its prescriptions for setting the right policy environment for reducing national greenhouse gas (GHG) emissions.

Initially, in 2007, there was good reason to believe that the CPRS would be both publicly and politically popular. Emissions trading regimes were consistent with the promoted benefits of neo-liberal governance where the uncertainties and risk-taking associated with the private market are embraced to enable socially productive outcomes (Bernstein 1996; O'Malley 2004). Current governance prescriptions imbue market mechanisms with particular efficacy. It is noteworthy, for example, that the Garnaut Review argued that an ETS could bear the primary burden for generating the economic motivation for spurring technological innovation and the efficiency gains necessary to reduce emissions to an acceptable level. Indeed, the Review maintained that policy measures additional to emissions trading could distort the regime and create unnecessary inefficiencies in Australia's path to a low carbon future. Further, it argued that:

> Programs and other regulatory interventions ... presuppose that government officials, academics or scientists have a better understanding of consumer preferences and technological opportunities than households and businesses. This is generally unlikely and cannot be guaranteed (Garnaut 2008a, p. 317).

Seen from this perspective, the failure of the CPRS is surprising since it drew explicitly on contemporary views of best practice regulation that eschews direct government intervention in favour of market mechanisms. Competition and consumer choice are seen to generate creative problem-solving by the private market. What was perhaps less surprising were the weaknesses within the CPRS itself that threatened its key policy goal, the reduction of GHG emissions in Australia to a level at least 5% below the 2000 level. Critics argued that, even had it passed into law, the CPRS may not have achieved the desired reduction in emissions, with carbon offsetting and subsidies to heavy polluters key aspects of the scheme (Australian Greens 2010; Daley and Edis 2010; Denniss 2009).

The failure of the CPRS in Australia points to the complex set of risk challenges that must be faced by policy initiatives set to tackle climate change. Significant challenges are posed by political risk, a risk that suggests limits to the effectiveness of neo-liberal governance strategies. The trajectory of the CPRS from its inception to its final defeat in parliament in February 2010 reveals the multi-faceted nature of political risk as composed of both economic and social components.

The argument in this chapter is that, to be successful, Australia's response to climate change must involve three separate ideal typical risks, each addressing different audiences and comprising inherently different dynamics. Policy measures must address all three to make a meaningful Australian contribution to the global challenge of reducing absolute emissions by 50–85% of 1990 levels by 2050 (IPCC 2007). The policy challenge clearly involves tangible reduction of emissions in order to avoid what I term the 'actuarial risks' of climate change. Yet policies must reduce emissions in a manner that reassures Australians about their safety and security (risks of a more socio-cultural nature) while enhancing political legitimacy and maintaining a vibrant economy (i.e. managing political risk) (for an extended discussion of these risk types see Haines 2011, Ch. 3). When considered in the light of actuarial, socio-cultural and political risks, an ETS as the central plank of GHG mitigation has merit in tackling the actuarial risk associated with rising emissions but contains particular weaknesses in surmounting socio-cultural and political risk challenges, weaknesses that require scrutiny.

Neo-liberalism, emissions trading and the limits of markets

At one level, ETSs seem well suited to the task of effective climate governance. An ETS fits well within stated government preferences for contemporary governance strategies that minimise direct governmental control (Grabosky 1994; Scott 2004; Shearing 1993). ETSs eschew direct measures and emphasise the creativity and problem-solving capacity of the private market in spurring innovation and efficiently allocating the resources needed to reduce emissions. Ross Garnaut's (2008a, p. 317) succinct heading, 'Don't pick winners. Fix market failures', characterises the essential vision of an ETS, namely that the costs of unwanted pollution are internalised and borne by those who pollute, thus providing the most effective means to reduce that pollution at the lowest cost.

Yet, there are weaknesses to market-based approaches in general and challenges to carbon trading markets in particular. In all markets, creativity can result in undesirable as well as desirable activity. The iatrogenic qualities of market systems result from attempts to 'game' market rules through various forms of creative compliance and regulatory arbitrage. Parallels can be drawn between the history of problems with financial markets and emerging problems in emissions trading. There is clear evidence of the incapacity of financial markets to control excessive risk-taking, and occasional outright fraud by stockbrokers, accountants and entrepreneurs intent on maximising private gain at public expense (Clarke *et al.* 2003; Furguson 2008; Gunningham 1991; Haines 2007; Reichman 1991). These same properties can be expected to emerge in emissions trading (Chan 2009; Jarman 2007; Schneider 2009). A particular problem with emissions trading markets stems from their closed nature. An ETS successful in reducing emissions will see fewer and fewer permits released (consistent with emission reduction targets), hence it cannot grow in the same manner as traditional stock markets (Baldwin 2008). The closed nature of ETSs means that strong political will is required to set sufficiently stringent emissions reduction targets – and to keep to them. The difficulty of setting sufficiently onerous demands on business was understood well before the inception of the CPRS and noted in the Garnaut Review (Garnaut 2008a, p. 314). The experience of both the trading system set up under the Kyoto protocol and the European ETS demonstrated that setting stringent targets and resisting the call for free allocation of permits or other exceptions to market discipline demanded a level of political courage that was lacking in those two systems (Baldwin 2008; Christoff 2006; Hepburn 2007; Oleschak and Springer 2007). The weakness in the design of an ETS that allows relief from market discipline has serious consequences for emissions abatement by undermining the incentive to generate the necessary new technology. The lack of market discipline in the international (Kyoto) ETS has failed to engender sufficient incentives for research in and development of new low carbon technology (Urge-Vorsatz *et al.* 2007). Even before the failure of the CPRS, Australians had reason to be wary of the promises made by politicians about the capacity of the ETS to deliver a significant reduction in emissions because of the need to commit to stringent reduction targets, and to deliver on them.

The particular characteristics of political risk explain why, despite these known problems, the CPRS was promoted as the heart of Australia's mitigation strategy. Political risk encompasses threats to political legitimacy, hence countering these threats plays a critical role in directing policy choice (Haines 2011). When understood through a Habermassian (Habermas 1979, 1989a, b) lens, the central dilemma facing governments of capitalist democracies in their management of political risk is how to maintain economic growth in the face of chronic instability (i.e. the booms and busts of the economic cycle) while simultaneously reassuring the electorate that it is well protected from threats and insecurities. These two components comprise threats to political legitimacy, hence either (or both) can exacerbate political risk.

It is important to note that in managing political risk, governments do not necessarily need to put in place effective measures to tackle actuarial risk. That is, they can manage political risk without tackling the problem of rising emissions. An ETS does not need to reduce emissions in order to reduce political risk. This is because political risk in the face of disaster can be reduced, at least in the short term, by keeping the economy moving and making people feel safe, without taking serious measures to reduce actuarial risk to the levels required (Haines 2009, 2011; Haines *et al.* 2008).

The possibility of providing sufficient reassurance without reducing emissions is enhanced by the particular actuarial risk at the heart of this problem – anthropogenic climate change. There are inherent uncertainties with scientific predictions about anthropogenic climate change and limited historical experience from which to learn, in contrast to experiences of a wide range of catastrophes from earthquakes to industrial disasters. Hence, anthropogenic climate change can be understood as a 'first-of-a-kind' disaster phenomenon, a form of disaster that provides particular policy challenges (Mitchell 1996).

Nonetheless, given the dire prognosis by scientists what is most puzzling about the progress in Australia to date is the lack of progress in climate policy-making. Given the size and significance of the problem an authoritative, even authoritarian, approach by governments to curbing greenhouse gases might have been expected, a response utilised against other threats such as terrorism and crime (Garland 2001; Loader and Walker 2007). Such measures have proved effective in assuaging public concern, even in the face of their ambiguous effects on solving the actuarial problem or reducing the risk of disaster. Yet the dominant responses by major parties in Australia at the federal level have been emissions trading and voluntary schemes, the latter approach well-honed by the Howard Coalition government (Parker 1999) and prominent in the Abbott-led Opposition in its most recent climate change policy document (Coalition 2010).

Understanding political risk and climate policy

Many commentators highlight the critical role political expediency has played in valorising ETSs as the central strategy in combating climate change (Hepburn 2007; Soleille 2006; van Asselt and Biermann 2007). European governments in particular have succeeded in promoting an ETS as central to reducing emissions. In doing so, they have been argued to have avoided the political challenges that accompany the imposition of a broad-based carbon tax or rigorous energy efficiency standards, measures likely to curtail 'essential' purchases made by voters and create an industry backlash (Hepburn 2007).

From the standpoint of political risk outlined above, however, the attraction of emissions trading can be seen to extend beyond an analysis of political expediency. Maintaining economic growth is a critical component in keeping political risk under control. Dampening demand and economic investment increases the likelihood of recession. This risks reducing the government revenue necessary for public spending while at the same time it invites accusations that a government has mismanaged the economy (Habermas 1989b; Haines 2011; Haines *et al.* 2008). A market-based mechanism to tackle global warming is appealing because it promises the lowest impact on the productive capacity of Australia's capitalist economy.

As Habermas (1989b) noted, political risk does not arise only when economic investment and growth is threatened, but when a regime appears incapable of reassuring citizens that it can shield them from the perils of the unknown. Reassuring citizens about their security is as critical to reducing political risk as keeping the economy burning at an even intensity (Habermas 1989b; Haines 2011; Haines *et al.* 2008).

Reducing political risk is not just about maintaining economic growth or about public reassurance, it is both. Succeeding in both arenas, though, is fraught. If a government can convince people that market-based strategies can not only satisfy economic needs but also protect from some impending risk to security, then it is likely to breathe more easily. Historically, however, the market has not proved particularly reassuring to people concerned with their own well-being (Habermas 1989b); indeed, there are arguments that markets themselves increase people's perception of insecurity (Garland 2001; Giddens 1991). This insecurity, exacerbated by particular threats, may increase demands for command and control regulation, where government mandates standards and enforces them through a dedicated regulatory agency. Government wielding direct authority through command and control provides greater reassurance to popular concerns about insecurity (Reason 1997; Sunstein 2005). Command and control ('law and order') responses reassure in a way that markets cannot (Garland 2001). With this in mind, the capacity for a well-designed ETS to actually make its way through the enactment process, reassuring people about their safety and security, meeting economists' demands to ignore special pleading and climate scientists' insistence upon swift and significant cuts to carbon emissions, appears remote.

Revisiting the nature of the climate problem

Why did the political elite consider that an ETS would reassure people that catastrophic levels of global warming can be averted? Where was the demand for the more decisive measures that often accompany environmental and human risks? It is worth considering in a little more detail the nature of this particular threat, the nature of the actuarial risk posed by climate change. As explained above, the risk is both chronic and crescive.[3] Climate change is not experienced with the visibility and immediacy of an aeroplane crash or an industrial disaster and, because of this, may not engender the same level of public anxiety and demand for political response (Boin and t'Hart 2003). Moreover, the cause of the problem, GHG emissions, is tightly embedded within contemporary modes of production (Beck 1992; Bulkeley 2001; Jackson 2009; Sutton and Haines 2003). These two reasons allow us to surmise why the crisis around climate change initially failed to galvanise demands for either strong command and control responses or for the Australian government to substantiate its claims that an ETS could deliver.

There is an additional challenge for climate policy-making, one that centres on the role of certainty and uncertainty in policy-making and in science. Successful policy development often involves claims that a particular scientific body of evidence means that policy-makers have 'no choice' but to act in a particular way (Bedsworth and Kastenberg 2002; Habermas 1989a). But science is premised on uncertainty, where today's knowledge is seen as contingent on tomorrow's experience and further research (Beyea and Berger 2001; Popper *et al.* 2005; Tilley 1980). Considerable scientific uncertainty remains in predicting the precise impact of elevated GHG emissions on our climate. The uncertainty of science, however, is quite distinct. The issue is to move to a closer and closer understanding of the real impact of humanity on the climate.

Scientific uncertainty is ripe for exploitation both politically and socio-culturally in the policy-making process (Bedsworth and Kastenberg 2002; Short and Clarke 1992; Wynne 2010; Ch. 12, this volume). Various coalitions of industry and governments have sought to discredit the science, pointing to the lack of certainty, and have successfully undermined the political motivation necessary for a strong regulatory response (Pearce 2007). This is despite the scientific uncertainty revolving more around predicting (from any raised concentrations of GHGs) exactly when, where and by how much temperatures will increase, sea levels rise or precipitation change. Scientific uncertainty is not whether there will be an impact, but how bad the impact will be. Indeed, recent revelations suggest a more dire scenario than previously thought (IPCC 2007).

It is important to note that there have been climate-related actuarial risk challenges that have generated persistent public debate but have not resulted in a market-based policy response. In Australia, the clearest manifestation of this is increasingly scarce water supplies. The responses by various levels of government to the water crisis affecting many cities and towns have involved water restrictions (a classic command and control response), building power-hungry desalination plants, transferring water from one region of scarcity to another and trucking in water. None of these are particularly climate-friendly. It is clear that a market-based solution to the problem of water scarcity in the residential sector[4] has failed to capture the attention of politicians wary that they will feel the wrath of the electorate should supplies of water run dry. The solution to the water crisis has been tighter command and control regulation, increased prices for water and assurance of supply through the construction of desalination plants across the country. The political crisis generated by domestic water scarcity is acute and hence leads to a quite distinct regulatory profile when compared with measures directly concerned with reduction in GHG emissions.

Risk management and the fate of the CPRS

It is important to identify the key advantage of ETSs in reducing emissions. Their strength lies in the identification of a target and the dissemination of permits that limit emissions to ensure that target is met. The cap determines a clear limit to absolute emission levels. Because of this, a well-designed ETS can spur an absolute, not just a relative, decline in emissions. That is, the cap set by the ETS, when combined with sufficient levels of compliance, can ensure that the overall emissions reduction target is met (Australian Government 2008). Theoretically, it can generate strong reduction of actuarial risk.

For this to occur, the reduction target must be robust, tight control retained over the number of permits released and few exemptions allowed. The Labor government failed to meet these challenges in the design of the CPRS. Both the target eventually set by the Australian government and the design of the scheme in terms of exemptions and assistance to industry fell well short of what was required, particularly in the short term. The initial target was set in a green paper, as a 5–15% reduction in GHG emissions by 2020 as measured by 2000 levels (Australian Government 2008). This was increased in May 2009 to 25% if agreement were struck in Copenhagen for reduction at that level. It was not. The target was low and the baseline was set at 2000, 10 years later than the 1990 level baseline favoured by the IPCC. The medium-term target was a little more promising at 60% below 2000 levels and within the IPCC range of 50–85%.

The design of the CPRS was also problematic, and became more so as the negotiation with the conservative Opposition proceeded. Despite known problems with the allocation of free permits (discussed above), they were to be provided to electricity generators under an Electricity Sector Adjustment Scheme (ESAS) with 228.7 million free permits to be provided at an approximate cost of $7.3 billion. The Emissions Intensive Trade Exposed Industry (EITE) also allowed up to 94.5% of the permits to be provided free. This was despite the fact that some of those supported by the ESAS actually had higher emissions than equivalent facilities offshore (e.g. aluminium). The effect of free permits would be to institutionalise high levels of emissions from these facilities within Australia (Daley and Edis 2010). Even within Australia, the free permits to the coal industry, for example, undermined movement away from mines with high emissions to those with lower emissions (Daley and Edis 2010).

Despite intense negotiation, the CPRS was defeated in parliament in February 2010. It failed as an exercise in political risk management. The defeat was partly caused by a change in Opposition leadership, with new Opposition leader Tony Abbott rejecting emissions trading as 'a big new tax' and returning to the voluntary policies of the Howard era, together (somewhat ironically) with greater public spending through the proposal for an Emissions Reduction Fund. The support of the Greens for the CPRS was lost because of the severely compromised nature of the scheme. Ultimately, the scheme failed due to all three aspects of risk: it failed to demonstrate actuarial risk reduction (i.e. reduction in GHG emissions), it failed to reassure an anxious public (i.e. it heightened socio-cultural risk) and it appeared too serious a risk to the political legitimacy of the Rudd Labor government. Threats by the resource sector and electricity generators about the negative impact on the Australian economy meant that the Labor government felt it was vulnerable both from socio-cultural and economic points of view. The conservative Opposition made good political use of both the economic vulnerabilities and public anxieties associated with the CPRS. The two pressures leading to heightened overall level of political risk were influential in the CPRS being considered a political liability and hence shelved until after the 2010 election. The Labor government went to the 2010 election promising it would create a Citizens' Assembly to develop a consensus on climate change.

Certainly, climate-related policies have been implemented, including a renewable energy target and enhanced efficiency standards. However, it is important to briefly consider the weaknesses in these alternative climate policies, particularly those targeted at increased energy efficiency and enhanced technology in the absence of an overall cap on emissions. They might be used to manage political risk effectively, maintaining economic growth and reassuring the population, yet also fail to reduce emissions. The difference between absolute and relative levels of emissions reduction needs careful attention. Requirements for increased efficiency are often accompanied by increased consumption, thus dampening the overall impact, a problem labelled 'technology rebound' (Jackson 2009). A useful example in Australia relates to domestic dwelling construction. Building efficiency levels are more stringent, but this stringency has been accompanied by increases in both the size of houses and the technology used in homes so that overall energy consumption of the average household has increased (ABS 2006; Giles 2007). The same is true for cars (Lucas 2008). Here again, economic growth is preserved and people feel reassured but emissions keep rising.

Conclusion

Difficult challenges lie ahead in institutionalising laws and policies capable of bringing about effective reduction of GHG emissions both internationally and within Australia. This chapter has argued that political risk must be understood and addressed if effective policies are to be implemented. Political risk can be managed by responding to the demands of industry for the conditions they consider necessary for ongoing economic investment and by putting in place policies that reassure an anxious public. Such measures may not include taking effective action to reduce actuarial risk, i.e. effective and timely reduction in GHG emissions.

The political attraction of emissions trading lies in its promise of preserving the investment and creativity of the market, responding to public anxiety and its theoretical capacity in reducing emissions. Yet the fate of the CPRS illuminated the failure of market strategies when faced with significant threats that exacerbated political risk to the incumbent Labor government. The first casualty was a weakening of resolve to tackle the significant actuarial risk, the second was a misreading of socio-cultural risk associated with market-based strategies and the third was a failure of political risk management. Overall, the history of the CPRS demonstrates that the actuarial risk challenge of climate change cannot be considered separately from political risk or socio-cultural anxieties. It is for this reason that the dialogue between natural and social scientists and between politicians, the public and industry becomes critical. The themes discussed in this book could not be more germane to addressing humanity's impact on the Earth's climate.

Endnotes

1 Australian Labor is the left-leaning political party within Australia.

2 The Coalition within Australia is composed of two political parties, the Liberals (the conservative right-wing party within Australia) and the Nationals (a political party largely representing rural, agricultural-based electorates). These two parties join to form the Coalition which stands in federal elections as a single prospective government. The Direct Action Plan was the policy document on climate change the Coalition took to the 2010 federal election.

3 Increasing, growing, expanding. Defined by Merriam Webster as 'marked by spontaneous gradual development'; definition available at <http://www.merriam-webster.com/dictionary/crescive>.

4 Market-based regimes on the allocation and trading of permits to utilise a scarce resource exist in the agricultural sector, particularly in the case of water. Overallocation of permits and chronic delays in the required scaling-back of those permits has meant that such measures have spectacularly failed to maintain critical river systems such as the Murray-Darling river system. It is a lesson from which proponents of an ETS might learn.

References

Australian Bureau of Statistics (ABS) (2006) Environmental impact of household energy use. <http://www.abs.gov.au/AUSSTATS/abs@.nsf/bb8db737e2af84b8ca2571780015701e/a300c2a2b4e0b91fca2571b000197552!OpenDocument> (accessed 19.05.08).

Australian Government (2008) Carbon Pollution Reduction Scheme Green Paper. <http://www.climatechange.gov.au/en/publications/cprs/green-paper/cprs-greenpaper.aspx> (accessed 28.07.10).

Australian Greens (2010) The Greens and emissions trading: your questions answered. 10 January. <http://christine-milne.greensmps.org.au/content/greens-and-emissions-trading-%E2%80%93-your-questions-answered#cantsupportCPRS> (accessed 18.11.10).

Baldwin R (2008) Regulation lite: emissions trading. *Regulation and Governance* **2**, 193–215.

Beck U (1992) *Risk Society: Towards a New Modernity.* Sage, London.

Bedsworth LW and Kastenberg WE (2002) Science and uncertainty in environmental regulation: insights from the evaluation of California's smog check program. *Science and Public Policy* **29**, 13–24.

Bernstein PL (1996) *Against the Gods: The Remarkable Story of Risk.* John Wiley and Sons, New York.

Beyea J and Berger D (2001) Scientific misconceptions among Daubert gatekeepers: the need for reform of expert review procedures. *Law and Contemporary Problems* **64**, 327–372.

Boin A and t'Hart P (2003) Public leadership in times of crisis: mission impossible? *Public Administration Review* **63**, 544–553.

Bulkeley H (2001) Governing climate change: the politics of risk society. *Transactions of the Institute of British Geographers* **26**, 430–447.

Chan M (2009) Lessons learned from the financial crisis: designing carbon markets for environmental effectiveness and stability. *Carbon and Climate Law Review* **3**, 152–160.

Christoff P (2006) Post Kyoto? Post Bush? Towards an effective 'climate coalition of the willing'. *International Affairs* **82**, 831–860.

Clarke F, Dean G and Oliver K (2003) *Corporate Collapse: Accounting, Regulatory and Ethical Failure*. Cambridge University Press, Cambridge.

Coalition (2010) Direct Action Plan: environment and climate change. <http://www.liberal.org.au/~/media/Files/Policies%20and%20Media/Environment/The%20Coalitions%20Direct%20Action%20Plan%20Policy.ashx> (accessed 20.08.10).

Daley J and Edis T (2010) Restructuring the Australian economy to emit less carbon: main report. Grattan Institute. April. <http://www.carbonneutral.com.au/grattan_report_restructuring_aust_economy_to_emit_less_carbon_april_2010.pdf> (accessed 15.09.10).

Denniss R (2009) Carbon plan helps big polluters and falls down in big flaws. *Canberra Times*. 20 November. <https://www.tai.org.au/index.php?q=node%2F19&pubid=710&act=display> (accessed 26.07.10).

Furguson N (2008) *The Ascent of Money: A Financial History of the World*. Penguin, New York.

Garland D (2001) *The Culture of Control: Crime and Social Order in Contemporary Society*. University of Chicago Press, Chicago.

Garnaut R (2008a) *The Garnaut Climate Change Review*. Cambridge University Press, Melbourne.

Garnaut R (2008b) Interim Report to the Commonwealth, state and territory governments of Australia. Garnaut Climate Change Review, 1–63. <http://www.garnautreview.org.au/domino/Web_Notes/Garnaut/garnautweb.nsf> (accessed 19.09.08).

Giddens A (1991) *Modernity and Self-identity: Self and Society in the Late Modern Age*. Polity Press, Cambridge.

Giles J (2007) Building for a cooler plant. *New Scientist* 25 July. <http://environment.newscientist.com/channel/earth/mg19526143.800> (accessed 19.05.08).

Grabosky P (1994) Green markets: environmental regulation by the private sector. *Law and Policy* **16**, 420–448.

Gunningham N (1991) Private ordering, self-regulation and futures markets: a comparative study of informal social control. *Law and Policy* **13**, 297–326.

Habermas J (1979) *Legitimation Crisis*. Heinemann, London.

Habermas J (1989a) Technology and science as ideology. In *Jurgen Habermas on Society and Politics: A Reader*. (Ed S Seidman) pp. 237–265. Beacon Press, Boston.

Habermas J (1989b) What does a crisis mean today? Legitimation problems in late capitalism. In *Jurgen Habermas on Society and Politics: A Reader*. (Ed S Seidman) pp. 266–283. Beacon Press, Boston.

Haines F (2007) Crime? What crime? Tales of the collapse of HIH. In *International Handbook of White-collar and Corporate Crime*. (Eds HN Pontell and G Geis) pp. 523–539. Springer, New York.

Haines F (2009) Vanquishing the enemy or civilizing the neighbour? Controlling the risks from hazardous industries. *Social and Legal Studies* **18**, 397–415.

Haines F (2011) *The Paradox of Regulation: What Regulation can Achieve and What it Cannot*. Edward Elgar, London.

Haines F, Sutton A and Platania-Phung C (2008) It's all about risk, isn't it? Science, politics, public opinion and regulatory reform. *Flinders Journal of Law Reform* **10**, 435–453.

Hepburn C (2007) Carbon trading: a review of the Kyoto mechanisms. *Annual Review of Environment and Resources* **32**, 375–393.

IPCC (Intergovernmental Panel on Climate Change) (2007) Summary for policy-makers of the synthesis report of the IPCC 4th assessment report. <http://www1.ipcc.ch/pdf/assessment-report/ar4/syr/ar4_syr_spm.pdf> (accessed 19.05.08).

Jackson T (2009) Prosperity without growth? Sustainable Development Commission, UK. <http://www.sd-commission.org.uk/publications.php?id=914> (accessed 01.07.09).

Jarman M (2007) *Climate Change*. Pluto Press/Oxfam GB, London.

Loader I and Walker N (2007) *Civilizing Security*. Cambridge University Press, Cambridge.

Lucas C (2008) Going nowhere in car efficiency. *The Age* 28 April. <http://www.theage.com.au/news/national/going-nowhere-fast-in-car-fuel-efficiency/2008/04/27/1209234656180.html> (accessed 19.05.08).

Mitchell JK (1996) Improving community responses to industrial disasters. In *The Long Road to Recovery: Community Responses to Industrial Disasters*. (Ed JK Mitchell) pp. 1–11. UN University Press, Tokyo.

Oleschak R and Springer U (2007) Measuring host country risk in CDM and JI projects: a composite indicator. *Climate Policy* 7, 470–487.

O'Malley P (2004) *Risk, Uncertainty and Government*. Glasshouse Press, London.

Parker C (1999) The greenhouse challenge: trivial pursuit? *Environmental and Planning Law Journal* **16**, 63–74.

Pearce G (2007) *High and Dry: John Howard, Climate Change and the Selling of Australia's Future*. Viking, Melbourne.

Popper SW, Lempert RJ and Bankes SC (2005) Shaping the future. *Scientific American* **292**, 66–71.

Reason J (1997) *Managing the Risks of Organisational Accidents*. Ashgate, London.

Reichman N (1991) Regulation risky business: dilemmas in security regulation. *Law and Policy* **13**, 263–346.

Schneider L (2009) Assessing the additionality of CDM projects: practical experiences and lessons learnt. *Climate Policy* **9**, 242–254.

Scott C (2004) Regulation in the age of governance: the rise of the post-regulatory state. In *The Politics of Regulation: Institutions and Regulatory Reforms for the Age of Governance*. (Eds J Jordan and DL Faur) pp. 145–174. Edward Elgar, London.

Shearing C (1993) A constitutive conception of regulation. In *Business Regulation and Australia's Future*. (Eds P Grabosky and J Braithwaite) pp. 67–73. Australian Institute of Criminology, Canberra.

Short JFJ and Clarke L (1992) Social organization and risk. In *Organizations, Uncertainties and Risk*. (Eds JFJ Short and L Clarke) pp. 309–321. Westview, Boulder.

Soleille S (2006) Greenhouse gas emission trading schemes: a new tool for the environmental regulator's kit. *Energy Policy* **34**, 1473–1477.

Sunstein C (2005) *Laws of Fear: Beyond the Precautionary Principle*. Cambridge University Press, Cambridge.

Sutton A and Haines F (2003) White collar and corporate crime. In *Crime and Justice: An Australian Textbook in Criminology*. (Eds A Goldsmith, M Israel and K Daly) pp. 141–158. Law Book Co., Sydney.

Tilley N (1980) Popper, positivism and ethnomethodology. *British Journal of Sociology* **31**, 28–45.

Urge-Vorsatz D, Keppel S and Mirasgedes S (2007) Appraisal of policy instruments for reducing buildings' CO_2 emissions. *Building Research and Innovation* **35**, 458–477.

van Asselt H and Biermann F (2007) European emissions trading and the international competitiveness of energy-intensive industries: a legal and political analysis of possible supporting measures. *Energy Policy* **35**, 497–506.

Wynne B (2010) Strange weather, again: climate science as political art. *Theory, Culture and Society* **27**, 289–305.

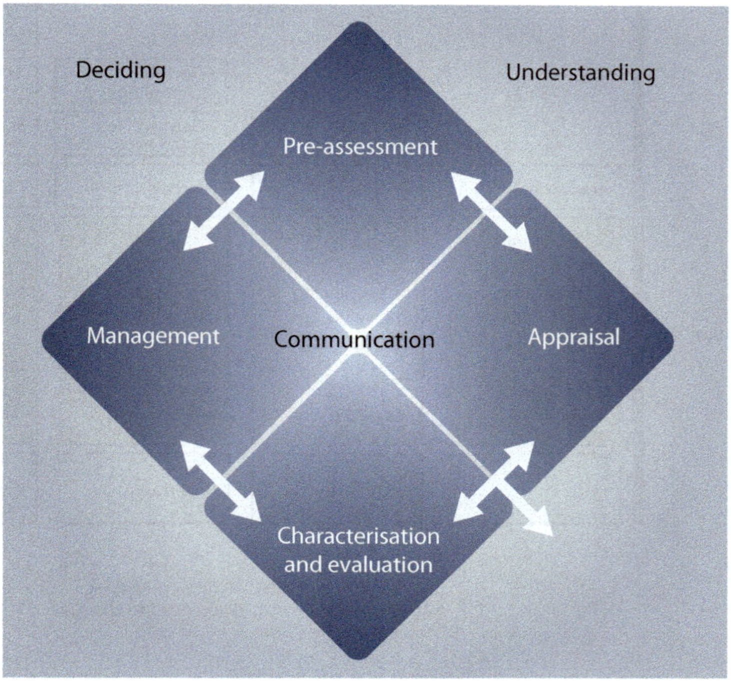

Figure 5.1: The risk governance framework.

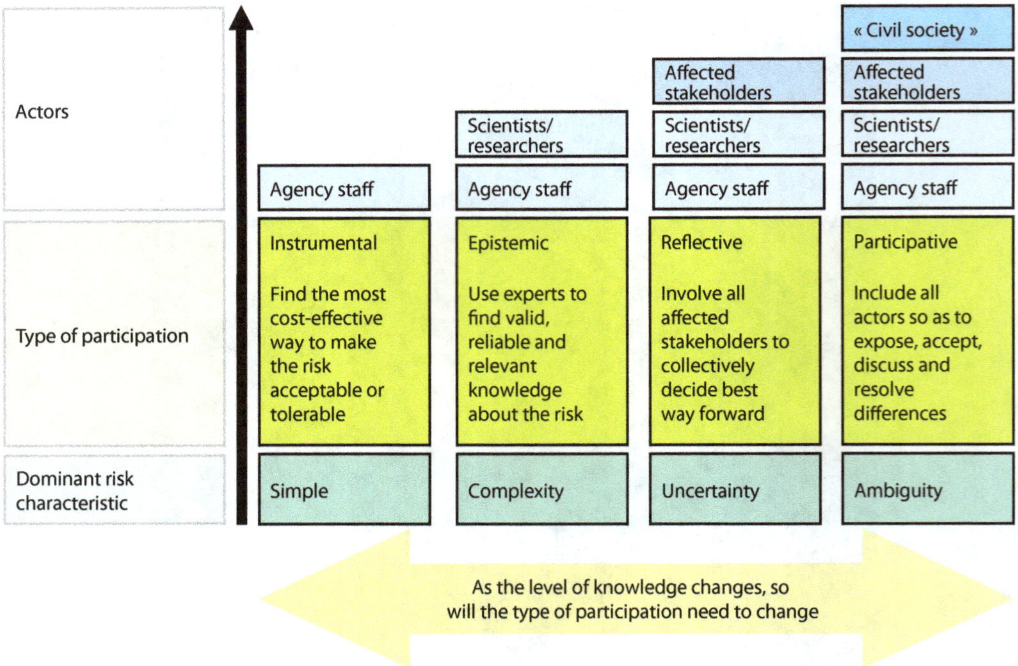

Figure 5.4: The risk escalator – a guide for inclusive risk governance.

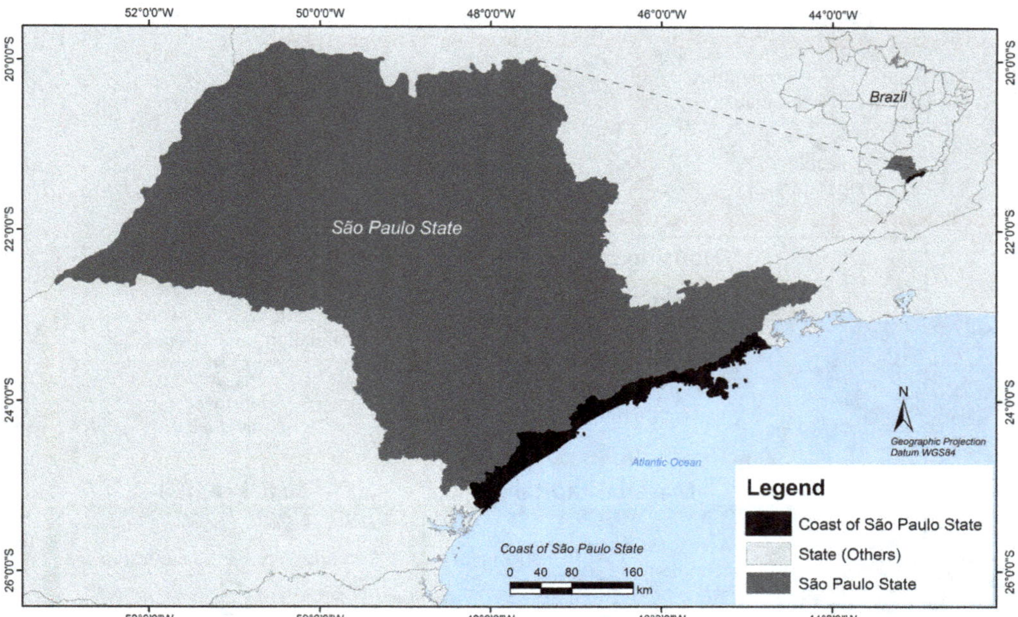

Figure 10.1: Coastal cities in São Paulo state, Brazil. Elaborated by Allan Iwama (2010).

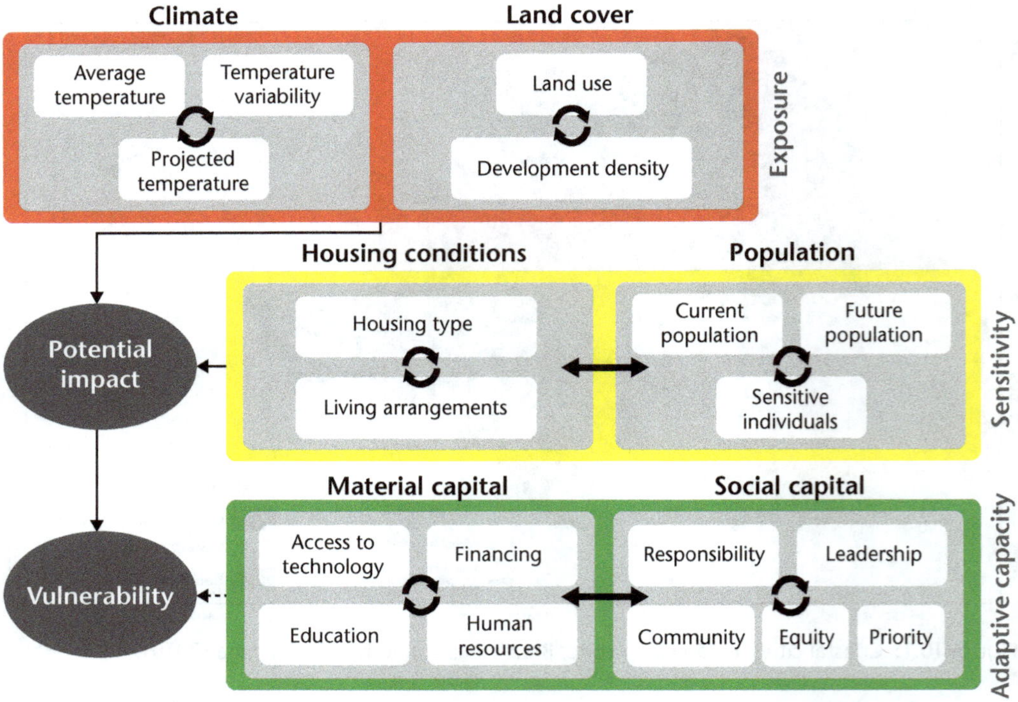

Figure 11.1: Conceptual model of the vulnerability of human health to extreme heat events. Exposure (red) is driven by interactions between the climate system and the landscape. Sensitivity (yellow) is a function of the characteristics of the exposed population and the conditions in which they live. Adaptive capacity (green) is a function of the material and social capital that can address potential impacts and ameliorate (Preston *et al.* 2008).

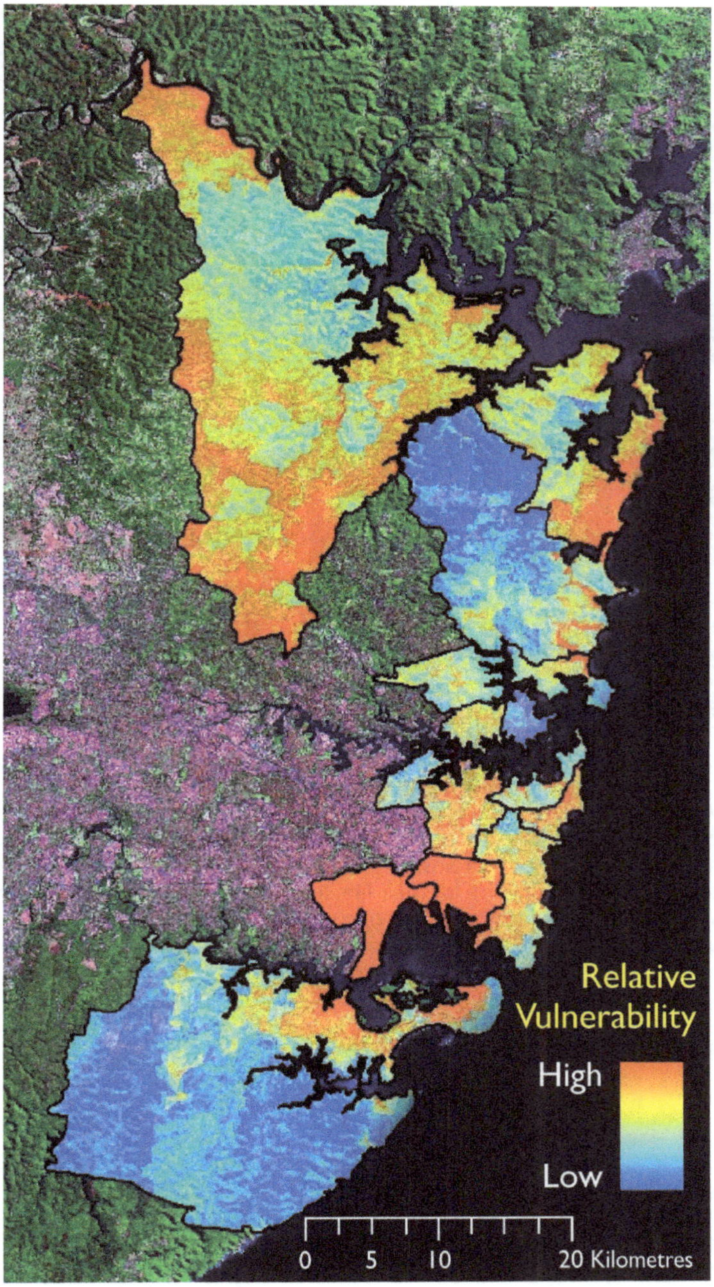

Figure 11.2: Map showing a composite of net vulnerability across the study region. Councils within the area were Botany Bay, Hornsby, Leichhardt, Manly, Mosman, North Sydney, Pittwater, Randwick, Rockdale, Sutherland, Sydney, Warringah, Waverley, Willoughby and Woollahra (Preston *et al.* 2008).

Figure 15.2: Four scenarios for the future of Makanya district, Tanzania (Enfors *et al.* 2008). The study argued that policies to reduce risk and support farmer innovation (e.g. micro-credit, knowledge exchanges) would be robust across the range of futures captured in the scenarios. 'Agricultural advance' describes a future dominated by a rapid market-oriented agricultural development. In 'Managing on the margins,' the dry climate makes farming-based livelihoods difficult and people struggle to provide for their families. 'Community cohesion' centres on collective action and small-scale improvements of current farming systems. 'Industry imperial' describes a future where smallholder farming loses importance as people start working in industries and factories

5 COMPLEXITY, UNCERTAINTY AND AMBIGUITY IN INCLUSIVE RISK GOVERNANCE

Ortwin Renn and Andreas Klinke

This chapter will conceptualise the role of complexity, uncertainty and ambiguity in risk governance. The main objective is to design and implement a systematic approach to organisational and policy learning in institutional settings that are conducive to resolving cognitive, evaluative and normative problems and conflicts of risks. We propose a risk governance model that augments the classical model of risk analysis (risk assessment, management, communication) by including steps of pre-estimation, interdisciplinary risk estimation, risk characterisation and evaluation and risk management as well as monitoring and control. This new risk governance model incorporates expert, stakeholder and public involvement as a core feature in the stage of communication and deliberation.

Introduction

Deciding the location of hazardous facilities, setting standards for chemicals, making decisions about clean-ups of contaminated land and regulating food and drugs as well as designing and enforcing safety limits have one element in common: these activities are collective endeavours to understand, assess and handle risks to human health and the environment. These attempts are based on two requirements. First, risk managers need sufficient knowledge about the potential impacts of the risk sources under investigation and the likely consequences of the different options to control these risks. Second, they need criteria to judge the desirability or undesirability of these consequences for the people affected and for the general public (Renn and Schweizer 2009; Horlick-Jones *et al.* 2007; Rowe and Frewer 2000). Criteria on desirability are reflections of social values such as good health, equity or efficient use of scarce resources. Both components – knowledge and values – are necessary for any decision-making process independent of the issue and the problem context.

Anticipating the consequences of human actions or events (knowledge) and evaluating the desirability and moral quality of these consequences (values) pose particular problems if the consequences are complex and uncertain and the values contested and controversial. Dealing with complex, uncertain and ambiguous outcomes often leads to the emergence of social conflict. Questions of how to deal with complex, uncertain and controversial risks demand procedures for dealing with risks that go beyond the conventional risk management routines. Numerous strategies to cope with this challenge have evolved. They include technocratic decision-making through the explicit involvement of expert committees, muddling through in a pluralist society, negotiated rule-making via stakeholder involvement, deliberative democracy or ignoring probabilistic information altogether (see reviews in Nelkin and Pollak 1979, 1980; Brooks 1984; Renn 2008) The main argument of this chapter is that risk management institutions need more adequate governance structures and procedures that enable them to integrate professional assessments (systematic knowledge), adequate institutional process (political legitimacy), responsible handling of public resources (efficiency) and public knowledge and perceptions (reflection on public values and preferences). These inputs require the involvement of several actors in the risk assessment and risk management process. The structures that evolve from the cooperation of various actors in all phases of the risk-handling process are subsumed in the term risk governance (IRGC 2005; Renn 2008). Hutter (2006, p. 215) characterises the move from governmental regulation to governance in the following manner:

> This decentring of the state involves a move from the public ownership and centralized control to privatized institutions and the encouragement of market competition. It also involves a move to a state reliance on new forms of fragmented regulation, involving the existing specialist regulatory agencies of state but increasingly self-regulating organizations, regimes of enforced self-regulation … and American-style independent regulatory agencies.

Governing choices in modern societies are seen as an interplay between government institutions, economic forces and civil society actors such as non-governmental organisations (NGOs). 'Risk governance' involves the translation of the substance and core principles of governance to the context of risk and risk-related decision-making (Hutter 2006). It includes, but also extends beyond, the three conventionally recognised elements of risk analysis (risk assessment, risk management and risk communication). It requires consideration of the legal, institutional, social and economic contexts in which a risk is evaluated, and the involvement of the actors and stakeholders who represent them. Risk governance looks at the complex web of actors, rules, conventions, processes and mechanisms concerned with how relevant risk information is collected, analysed and communicated, and how management decisions are taken.

Based on our previous work on risk governance and risk evaluation (Klinke and Renn 2001, 2002, 2010; Klinke *et al.* 2006; Renn 2008; Renn *et al.* 2011), this chapter expands on a normative-analytical model of a risk governance process that interlinks diverse actors and their claims, elaborates the institutional means to process diverse inputs and discusses the prospects and implications for adaptive capacity. The focus will be on collectively binding risk management decisions rather than on private risk management decisions. Such collective decisions are not only a product of government policy-making but are joint products by a wide variety of actors including scientists, the private sector, civil society and government agencies.

In this chapter we will analyse the major characteristics of risk knowledge then address major functions of the risk governance process: pre-estimation, interdisciplinary risk estimation (including scientific risk assessment and concern assessment), risk characterisation and risk evaluation as well as risk management including decision-making and implementation. We will explicate the design of an effective and fair institutional arrangement, including four different forms of public and stakeholder involvement, to cope with the challenges raised by the three risk characteristics. We conclude with some general lessons for risk governance.

Three characteristics of risk knowledge

Integrative risk governance is expected to address challenges raised by three risk characteristics that result from a lack of knowledge and/or competing knowledge claims about the risk problem. Transboundary and collectively relevant risk problems such as global environmental threats (climate change, loss of biological diversity, chemical pollution etc.), new and/or large-scale technologies (nanotechnology, biotechnology, offshore oil production etc.), food security or pandemics are all characterised by limited and sometimes controversial knowledge with respect to their risk properties and their implications (Horlick-Jones and Sime 2004). The three characteristics are complexity, scientific uncertainty and socio-political ambiguity (Klinke and Renn 2002, 2010; Klinke *et al.* 2006; Renn 2008).

Complexity

Complexity refers to the difficulty of identifying and quantifying causal links between a multitude of potential candidates and specific adverse effects (Lewin 1992; Underdal 2009). A crucial aspect concerns the applicability of probabilistic risk assessment techniques. If the chain of events between a cause and an effect follows a linear relationship (e.g. car accidents, an overdose of pharmaceutical products), simple statistical models are sufficient to calculate the probabilities of harm. Such simple relationships may still be associated with high uncertainty, for example if only few data are available or the effect is stochastic by its own nature. Sophisticated models of probabilistic inferences are required if the relationship between cause and effects becomes more complex (Renn and Walker 2008). The nature of this difficulty may be traced back to interactive effects among these candidates (synergisms and antagonisms, positive and negative feedback loops), long delays between cause and effect, inter-individual variation, intervening variables and others. It is precisely these complexities that make sophisticated scientific investigations necessary since the cause–effect relationship is neither obvious nor directly observable. Non-linear response functions may also result from feedback loops that constitute a complex web of intervening variables. Complexity requires therefore sensitivity to non-linear transitions as well as to scale (on different levels). It also needs to take into account a multitude of exposure pathways and the composite effects of other agents present in the exposure situation. Examples of highly complex risk include sophisticated chemical facilities, synergistic effects of potentially eco-toxic substances on the environment, failure risk of large interconnected infrastructures and risks of critical loads to sensitive ecosystems.

Scientific uncertainty

Scientific uncertainty relates to the limitation or absence of scientific knowledge (data, information) that makes it difficult to assess the probability and possible outcomes of undesired effects (Aven and Renn 2009; Filar and Haurie 2010; Rosa 1997). It most often results from an incomplete or inadequate reduction of complexity in modelling cause–effect chains (Marti *et al.* 2010). Whether the world is inherently uncertain is a philosophical question that we do not pursue here. It is essential to acknowledge, in the context of risk assessment, that human knowledge is always incomplete and selective and thus contingent upon uncertain assumptions, assertions and predictions (Functowicz and Ravetz 1992; Laudan 1996; Renn 2008). It is obvious that the modelled probability distributions within a numerical relational system can represent only an approximation of the empirical relational system that helps elucidate and predict uncertain events. It therefore seems prudent to include additional aspects of uncertainty (van Asselt 2000). Although there is no consensus in the literature on the best means of disaggregating uncertainties, the following categories appear to be an appropriate means of distinguishing between the key components of uncertainty.

- *Variability* refers to the different vulnerability of targets, such as the divergence of individual responses to identical stimuli among individual targets within a relevant population such as humans, animals, plants and landscapes.

- *Inferential effects* relate to systematic and random errors in modelling, including problems of extrapolating or deducing inferences from small statistical samples, animal data or experimental data onto humans or from large doses to small doses etc. These are usually expressed through statistical confidence intervals.
- *Indeterminacy* results from genuine stochastic relationship between cause and effects, apparently non-causal or non-cyclical random events, or badly understood non-linear chaotic relationships.
- *System boundaries* allude to uncertainties stemming from restricted models and the need to focus on a limited number of variables and parameters.
- *Ignorance* means the lack of knowledge about the probability of occurrence of a damaging event and about its possible consequences.

The first two components of uncertainty qualify as statistically quantifiable uncertainty and therefore can be reduced by improving existing knowledge, applying standard statistical instruments such as Monte Carlo simulation and estimating random errors within an empirically proven distribution. The last three components represent genuine uncertainty components and can be characterised to some extent by using scientific approaches, but cannot be completely resolved. The validity of the end results is questionable and, for risk management purposes, additional information is needed, such as a subjective confidence level in risk estimates, potential alternative pathways of cause–effect relationships, ranges of reasonable estimates, maximum loss scenarios and others. Examples of high uncertainty include many natural disasters, such as earthquakes, possible health effects of mass pollutants below the threshold of statistical significance, regional impacts of global climate change, environmental degradation due to the loss of biological diversity, cumulative effects of various environmental hazards, long-term effects of introducing genetically modified species into the natural environment and the occurrence of pandemics (e.g. SARS, avian flu) caused by viruses characterised by a rapid rate of mutation

Socio-political ambiguity

While more and better data and information may reduce scientific uncertainty, more knowledge does not necessarily reduce ambiguity. Ambiguity indicates a situation of ambivalence in which different and sometimes divergent streams of thinking and interpretation about the same risk phenomena and their circumstances are apparent (Feldman 1989; Zahariadis 2003). We distinguish between interpretative and normative ambiguity, which both relate to divergent or contested perspectives on the justification, severity or wider 'meanings' associated with a given threat (Stirling 2003; Renn 2008).

Interpretative ambiguity denotes the variability of (legitimate) interpretations based on identical observations or data assessment results, e.g. an adverse or non-adverse effect. Variability of interpretation, however, is not restricted to expert dissent. Laypeople's perception of risk often differs from expert judgements because it is related to qualitative risk characteristics such as familiarity, personal or institutional control, assignment of blame and so on. In contemporary pluralist societies diversity of risk perspectives within and between social groups is generally fostered by divergent value preferences, variations in interests and very few, if any, universally applicable moral principles; all the more, if risk problems are complex and uncertain. Examples of high interpretative ambiguity include low-dose non-ionising radiation, low concentrations of genotoxic substances or the impacts of alien species on natural environments. (How far do we go back in time to determine what species is alien and what domestic?)

That leads us to the aspect of normative ambiguity. This alludes to different concepts of what can be regarded as tolerable, for example referring to ethics, quality of life parameters, and distribution of risks and benefits. Ambiguity emerges where the problem lies in agreeing on the appropriate values, priorities, assumptions or boundaries to be applied to the definition of possible outcomes. Normative ambiguities can be associated, for example, with exposure to noise, aquaculture in sensitive areas, pre-natal genetic screening and genetically modified food. In these cases, science is very familiar with the risks and there is little uncertainty and interpretative ambiguity about cause–effect relationships. Yet there is considerable debate on whether the application is tolerable. Another example may be the use of phthalates in toys. All analysts are aware that the substance is potentially carcinogenic but, given the known exposure and the dose-response functions, there is hardly any possibility for young children to be negatively affected (Wilkinson and Lamb 1999). Yet the mere idea of having a carcinogenic substance in children's toys has incited a fierce debate about the tolerability of such an ingredient in rubber toys.

Most risks are characterised by a mixture of complexity, uncertainty and ambiguity. Passive smoking may be a good example of low complexity and uncertainty, but high ambiguity. Nuclear energy may be a good candidate for high complexity and high ambiguity, but relatively little uncertainty. The massive emission of aerosols to the atmosphere to combat the effects of greenhouse gases might be cited as an example of high complexity, uncertainty and ambiguity.

Towards an inclusive risk governance model

The ability of risk governance institutions to cope with complex, uncertain and ambiguous consequences and implications has become a central concern to scientists and practitioners alike. In 2005, the International Risk Governance Council suggested a process model of risk governance (IRGC 2005; Renn 2008). This framework structures the risk governance process in four phases: pre-assessment, appraisal, characterisation and evaluation, and risk management. Communication is conceptualised as a constant companion to all four phases of the risk governance cycle. The framework's risk process, or risk-handling chain, is illustrated in Figure 5.1 (in colour section).

Since its publication in 2005, the IRGC risk governance framework has been applied to diverse risk governance issues in various case studies. Publications of these case studies are available on IRGC's homepage (www.irgc.org/Publications.html). The case studies deal with emerging risks such as air quality, bioenergy, carbon capture and storage, critical infrastructure, nanotechnology, pollination services and synthetic biology. The IRGC has commissioned several case studies as tests of the applicability, efficacy and practicability of the risk governance framework (Renn and Walker 2008). The applications have shown that the framework can be used as broad conceptual guidance on the critical elements of the risk governance process. To date, the IRGC risk framework has been discussed and partially applied to a number of institutions and organisations, most prominently including the European Food Safety Authority (Voss and Wendler 2009) and the Health Council of the Netherlands (Health Council 2009). Reports using the framework have been given by the German Occupational Health and Safety Committee (Bender 2008), the International Occupational Safety Association (Radandt *et al.* 2008), the UK Treasury (UK Treasury 2005), the US EPA (2009) and several private organisations. The framework was also applied for strategic risk management by the US Joint Chiefs of Staff (Rouse 2011). It has been used for major military operations and has, according to the source, improved the risk management process considerably.

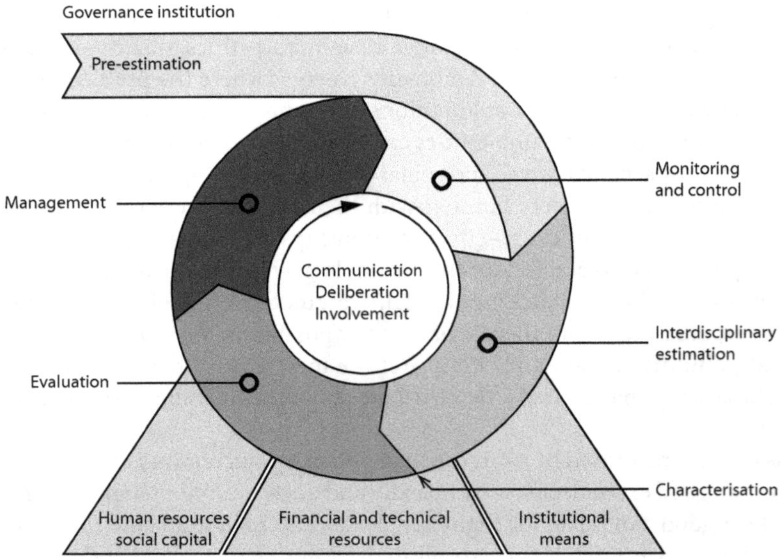

Figure 5.2: Adaptive and integrative risk governance model.

The framework was primarily developed to deal with technological risks. It has been criticised as overstating the demarcation line between assessment and management, as being too rigid in its phasing of the governance process and in being not specific enough on stakeholder involvement and participation (Renn and Walker 2008; van Asselt 2005). For the purpose of developing a more adaptive and inclusive version of the IRGC framework, Klinke and Renn (in press) and Renn *et al.* (2011) suggest a slightly modified version (Fig. 5.2).

The modified framework consists of several steps: pre-estimation, interdisciplinary risk estimation, risk characterisation, risk evaluation and risk management. These are all related to the ability and capacity of risk governance institutions to use resources effectively (see Fig. 5.2). Appropriate resources include institutional and financial means as well as social capital (e.g. strong institutional mechanisms and configurations, transparent decision-making, allocation of decision-making authority, formal and informal networks that promote collective risk-handling, education), technical resources (e.g. databases, computer software and hardware) and human resources (e.g. skills, knowledge, expertise, epistemic communities). Hence, the adequate involvement of experts, stakeholders and the public in the risk governance process is crucial in producing and conveying adaptive and integrative capacity in risk governance institutions (Pelling *et al.* 2008). Since the social acceptance of any response of risk governance to risk problems associated with complexity, uncertainty and/or ambiguity is critical, risk-handling and response strategies need to be flexible and the risk management approaches need to be iterative and inclusionary.

Pre-estimation

Risks are not straightforwardly objective phenomena. They are also mental constructions that reflect how people perceive uncertain phenomena and the ways in which their interpretations and responses are determined by social, political, economic and cultural contexts and judgements (Luhman 1993; OECD 2003; IRGC 2005). The introduction of risk as a mental construct depends on the presumption that human action can prevent harm. The conceptualisation of risk as a mental construct has major implications for how risk is considered. Risks are created and selected by human actors. What counts as a risk to some people may be perceived by others as a destiny explained by religion or as an opportunity. Although societies have gained experience and collective knowledge of the potential impacts of events and activities, we cannot anticipate all potential scenarios and be worried about the many possible consequences of a proposed activity or an expected event. By the same token, it is impossible to include all possible options for intervention. Therefore, societies have been selective in what they have chosen to consider worth addressing and what to ignore.

The insight that risks are not objective entities that need only be discovered, but are mental constructs of how people select signals from their environment in order to be better prepared if a hazard strikes, leads to the necessity to establish (culture-sensitive) institutions for early warning. Many such institutions exist, particularly on the national level. However, there is no adequate international mechanism for the detection and early warning of some globally relevant potential future harm (e.g. global environmental threats, global food security) in spite of the fact that some specialised organisations (e.g. International Atomic Energy Agency, Pacific Tsunami Warning Center, National Hurricane Center in Miami, Center for Disease Control in Denver) have been established to monitor specific hazards and to detect hints of future problems on an international scale.

A systematic review of the stages in pre-estimation would start with screening, as an exploration of a large array of actions and problems looking for those with a specific risk-related feature. It is important to explore what major political and societal actors such as governments, companies, epistemic communities, NFOs and the general public identify as risks and what types of problems they label as problems associated with risk and uncertainty. This is called framing, and it specifies how society and politics rely on schemes of selection and interpretation to understand and respond to the phenomena which are socially constructed as relevant risk topics (Kahneman and Tversky 2000; Reese *et al.* 2003). Interpretations of risk experience depend on the frames of reference (Daft and Weick 1984). The process of framing corresponds with a multi-actor and multi-objective governance structure since governmental authorities (national, supranational and international agencies), risk and opportunity producers (e.g. industry), those affected by risks and opportunities (e.g. consumer organisations, environmental groups) and interested bystanders (e.g. the media or an intellectual elite) are all involved and often in conflict over the appropriate frame within which to conceptualise the problem. What counts as risk may vary among these actor groups. Whether an overlapping consensus evolves about what requires consideration as a relevant risk depends on the legitimacy of the selection rule. For example, the risks and benefits of biomass conversion for energy purposes can be seen within the frame of energy security, national independence, climate protection or economic development opportunities for rural areas. Depending on the frame, different types of risks and benefits may emerge. Furthermore, some benefits within one frame (e.g. national independence) may be a risk within another frame (e.g. economic opportunities for developing countries).

Interdisciplinary risk estimation

For politics and society to come to reasonable decisions about risks in the public interest, it is not enough to consider only the results of scientific risk assessment. To understand the concerns of affected stakeholders, information about risk perceptions and the further implications of the direct consequences of a risk is needed and should be taken into account by risk management.

Interdisciplinary risk estimation thus includes scientific assessment of risks to human health and the environment and assessment of related concerns as well as social and economic implications (IRGC 2005; Renn and Walker 2007). The interdisciplinary estimation process should be clearly dominated by scientific analyses but, in contrast to traditional risk regulation models, the scientific process should include the natural and technical as well as the social sciences, including economics. The interdisciplinary risk estimation comprises two stages:

1 risk assessment – experts from the natural and technical sciences produce the best estimate of potential physical harm from a risk source;
2 concern assessment – experts from the social sciences, including economics, identify and analyse the issues that individuals or society as a whole link to a certain risk. The repertoire of the social sciences, such as survey methods, focus groups, econometric analysis, macro-economic modelling or structured hearings with stakeholders may be used.

There are different approaches and proposals regarding how best to address the issue of interdisciplinary risk estimation. The German Advisory Council on Global Change (WBGU) developed eight criteria to characterise risks beyond the established assessment criteria (Klinke and Renn 2002; WBGU 2000). Some of the criteria have been used by different risk agencies or risk estimation processes (e.g. HSE 2001). The criteria include:

- extent of damage – adverse effects in natural units, e.g. death, injury, production loss;
- probability of occurrence – estimate of relative frequency, which can be discrete or continuous;
- incertitude – how we take account of uncertainty in knowledge, in modelling complex systems or in predictability in assessing a risk;
- ubiquity – geographical dispersal of damage;
- persistence – how long the damage will last;
- reversibility – whether the damage can be reversed;
- delay effects – latency between initial event and actual damage;
- potential for mobilisation – the broad social impact, e.g. whether the risk will generate social conflict:
  - inequity and injustice associated with the distribution of risks and benefits over time, space and social status;
  - psychological stress and discomfort associated with the risk or risk source (as measured by psychometric scales);
  - potential for social conflict and mobilisation (degree of political or public pressure on risk regulatory agencies);
  - spill-over effects that are likely to be expected when highly symbolic losses have repercussions on other fields such as financial markets or loss of credibility in management institutions.

These four sub-criteria of the last category reflect many factors that have been proven to influence risk perception. The appraisal guidance published by the UK Treasury Department in 2005 recommended a risk estimation procedure that is similar to our proposal. It includes the results of risk assessment and direct input from data on public perception and the assessment of social concerns (HM Treasury 2005).

Risk evaluation

A heavily disputed task in the risk governance process relates to the procedure of how to judge a given risk and justify an evaluation about its societal acceptability or tolerability (see Fig. 5.2). In many approaches, risks are ranked and prioritised based on a combination of probability

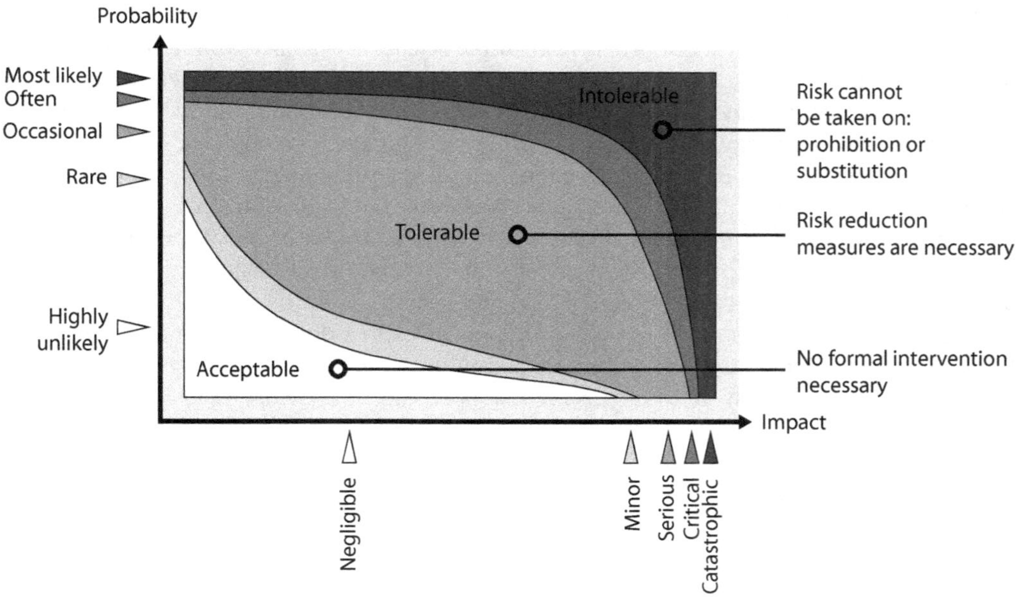

Figure 5.3: Risk areas.

(how likely it is that the risk will occur) and impact (the consequences, if the risk does occur). In the so-called traffic light model (Fig. 5.3), risks are located in the diagram of probability versus expected consequences and three areas are identified – green, yellow and red (Klinke and Renn 2002; Renn 2008). A risk falls into the green area if the occurrence is highly unlikely and the impact is negligible. No further formal intervention is necessary. A risk is seen as tolerable when serious impacts might occur occasionally (yellow area). The benefits are worth the risk, but risk reduction measures are necessary. A risk is viewed as intolerable when the occurrence of catastrophic impacts is most likely (red area). Possible negative consequences of the risk are so catastrophic that it cannot be tolerated in spite of potential benefits.

Drawing the lines between 'acceptable', 'tolerable' and 'intolerable' risk is one of the most controversial tasks in the risk governance process. The UK Health and Safety Executive developed a procedure for chemical risks based on risk–risk comparisons (Löfstedt 1997). Some Swiss cantons such as Basle County experimented with round tables as a means to reach consensus on drawing the two demarcation lines, whereby participants in the round table represented industry, administrators, county officials, environmentalists and neighbourhood groups. Irrespective of the selected means, the judgement on acceptability or tolerability is contingent on making use of a variety of different knowledge sources. One needs to include the data and insights resulting from the risk assessment activity, and additional data from the concern assessment.

Attempting to draw fixed and precise lines in the 'traffic light model' has been criticised as simplistic. To meet this objection, we have introduced gradual transition zones to allow more differentiation in the evaluation of risks (Klinke and Renn, in press). If a given risk based on the risk assessment results cannot definitely be classified as 'acceptable', but it is plausible that the occurrence is rather rare and the impact is rather minor, the risk would be assigned to the transition zone between 'acceptable' and 'tolerable'. Furthermore, we would locate a given risk in the transition zone between 'tolerable' and 'intolerable', if the risk cannot clearly be characterised as 'intolerable' and, at the same time, the benefits to society are seen at least partially worthwhile.

Risk evaluations in general rely on causal and principal beliefs as well as worldviews (Goldstein and Keohane 1993). Causal beliefs refer to the scientific evidence from risk assessment, whether, how and to what extent the hazard potential causes harm to the environment or to human health. This dimension emphasises cause–effect relations and provides factual guidance on which strategy is appropriate to achieve the goal of risk avoidance or reduction.

However, the question of what is safe enough implies a moral judgement about the acceptability of risk and the tolerable burden that risk producers can impose on others. The results of the concern assessment can provide insights into what kind of associations are present and which moral judgements people would prefer if able to choose. Of major importance is the perception of just or unjust distribution of risks and benefits. How these moral judgements are made and justified largely depends on cultural values and worldviews. They affect personal thinking and evaluation strategies and are shaped by collectively shared ontological and ethical convictions. The selection of strategies for risk-handling is therefore understandable only within the context of broader worldviews. Hence, society can never derive acceptability or tolerability from looking at the evidence alone. Likewise, evidence is essential if we are to know whether a value has been violated, or to what degree.

In cases where there is unanimous agreement about the underlying values and even the threshold of what is regarded as acceptable or tolerable (interpretive ambiguity), evidence in the form of risk estimates may be sufficient to locate the risk within the risk area diagram. A judgement can then best be made by those who have most expertise in risk and concern assessments, in which case it makes sense to place this task within the domain of risk estimation by experts. However, with respect to values and evidence we can distinguish three dimensions of socio-political ambiguity that make it difficult to simply delegate risk estimation to experts. Interpretive and normative ambiguity have already been discussed. These refer to ambiguity in relation to the interpretation of evidence and in relation to the underlying values that should be considered by risk managers. If there is disagreement about the interpretation of results, for example whether a change in the environment due to human intervention (e.g. adding nutrients to a lake) is an adverse effect, expertise alone does not resolve the issue. In this case, an epistemic or interpretative discourse among those who are knowledgeable about the subject is necessary to collect evidence on impacts and arguments about their evaluation. The US National Academy has labelled such a discourse as analytic-deliberative (Stern and Fineberg 1996). Even more deliberation is needed for resolving normative ambiguity. If the underlying values of what could be interpreted as acceptable or tolerable are disputed, while the evidence of what is at stake is clear and non-controversial, the judgement needs to be based on a discourse about values and their implications. Such a discourse should be part of risk management by the legitimate decision-makers.

A third dimension evolves when both interpretive and normative ambiguity are present, i.e. where both the evidence and the values are controversial. This would imply that assessment should involve an activity to find common ground for characterising and qualifying the evidence and risk management needs to establish agreement about the appropriate values and their application. A good example may be the interpretative and normative implications of global climate change. The Intergovernmental Panel on Climate Change (IPCC) has made considerable efforts to articulate a common characterisation of climatic risks and their uncertainties. Given the remaining uncertainties and the complexities of the causal relationships between greenhouse gases and climate change, both contested evidence and conflicting values affect whether governments prioritise prevention, mitigation or adaptation (Keeney and McDaniels 2001).

Since the last case includes both issues of the other two, the process of judging the tolerability and acceptability of a risk can be structured into two distinct components: risk characterisation and risk evaluation (IRGC 2005). The first step, risk characterisation, determines the evidence-based component for making the necessary judgement on the tolerability and/or acceptability of a risk. The step of risk evaluation determines the value-based component for making that judgement. The separation of evidence and values underlying the distinction between characterisation and evaluation is functional and not necessarily organisational (Renn 2008). Since risk characterisation and evaluation are closely linked and interdependent, it may even be wise to perform the two steps simultaneously in a joint effort by both risk assessment experts and risk management decision-makers (Frewer and Salter 2007). The US regulatory system tends to favour an organisational combination of characterisation and evaluation, while European risk management tends to maintain the organisational separation, e.g. in the food area (Löfstedt and Vogel 2001).

Risk management

Risk management starts by reviewing all relevant data and information gathered in the previous steps of interdisciplinary risk estimation, characterisation and risk evaluation. The systematic analysis of risk management options focuses on still-tolerable risks (yellow area) and those where tolerability is disputed (light green and orange transition zones). The other cases (green and red areas) are fairly easy to deal with. Intolerable risks demand prevention and prohibition strategies as a means of replacing the hazardous activity with another activity leading to identical or similar benefits. The management of acceptable risks is left to private actors (civil society, the economy). They may initiate additional and voluntary risk reduction measures or seek insurance for possible but minor or negligible losses. If risks are classified as tolerable or if there is a dispute on whether they are in the transition zones of tolerability, public risk management needs to design and implement actions that make these risks either acceptable or tolerable, by introducing reduction strategies. This task can be described in terms of classic decision theory (Aven and Vinnem 2007; Klinke and Renn 2010):

- identification and generation of generic risk management options;
- assessment of risk management options with respect to predefined criteria;
- evaluation of risk management options;
- selection of appropriate risk management options;
- implementation of risk management options;
- monitoring and control of option performance.

Meeting the different challenges raised by complexity, scientific uncertainty and sociopolitical ambiguity, it is possible to design general strategies for risk management that can be applied to four distinct categories of risk problems, thus simplifying the process step of risk management mentioned above (Klinke and Renn 2002).

The first category refers to linear risk problems: they are characterised as having low scores on the dimensions of complexity, uncertainty and ambiguity. They can be addressed by linear risk management because they are normally easy to assess and quantify. Routine risk-handling within risk assessment agencies and regulatory institutions is appropriate for this category, since the risk problems are well known, sufficient knowledge of key parameters is available and there are no major controversies about causes and effects or conflicting values. The management includes risk–benefit analysis, risk–risk comparisons or other instruments of balancing pros and cons.

If risks are ranked high on complexity but rather low on uncertainty and ambiguity, they require the systematic involvement and deliberation of experts representing the relevant epistemic communities to produce the most accurate estimate of the complex relationships. It does not make much sense to integrate public concerns, perceptions or any other social aspects for resolving complexity unless specific knowledge from the concern assessment helps to untangle complexity. Complex risk problems therefore demand the risk-informed management that can be offered by scientists and experts applying methods of expanded risk assessment, determining quantitative safety goals, consistently using cost-effectiveness methods and monitoring and evaluating outcomes.

Risk problems that are characterised by high uncertainty but low ambiguity require precaution-based management. Since sufficient scientific certainty is either unavailable or unattainable, expanded knowledge acquisition may help to reduce uncertainty and thus move the risk problem back to first stage of handling complexity. If, however, uncertainty cannot be reduced by additional knowledge, risk management should foster and enhance precautionary and resilience-building strategies and decrease vulnerabilities in order to avoid irreversible effects. Appropriate instruments include containment, diversification, monitoring and substitution.

Finally, if risk problems are ranked high on ambiguity (regardless of whether they are low or high on uncertainty), discourse-based management is required. This demands participative processing. It includes the need to involve major stakeholders as well as the affected public. The goal of discourse-based risk management is to produce a collective understanding among all stakeholders and concerned public on interpretative ambiguity or to find legitimate procedures of justifying collectively binding decisions on acceptability and tolerability. It is important that a consensus or a compromise be achieved between those who believe that the risk is worth taking (perhaps because of self-interest) and those who believe that the pending consequences do not justify the potential benefits of the risky activity or technology.

Communication, deliberation and involvement of non-governmental actors

The effectiveness and legitimacy of the risk governance process depends on the capability of management agencies to resolve complexity, characterise uncertainty and handle ambiguity by means of communication and deliberation.

Instrumental processing involving governmental actors

Dealing with linear risk issues, which are associated with low scores of complexity, scientific uncertainty and socio-political ambiguity, requires hardly any changes to conventional public policy-making. The data and information on such linear (routine) risk problems are provided by statistical analysis, law or statutory requirements determine the general and specific objectives, and public policy ensures that all necessary measures of safety and control are implemented and enforced. Traditional cost–benefit analyses including effectiveness and efficiency criteria are the instruments of political choice for finding the right balance between under- and overregulation of risk-related activities and goods. Monitoring the issue area is important as insurance against unexpected consequences. For this reason, linear risk issues can well be handled by an instrumental involvement of departmental and agency staff and enforcement personnel of state-run governance institutions. The aim is to find the most cost-effective method for a desired regulation level. If necessary, stakeholders may be included in the deliberations as they have information and know-how that may provide useful hints for being more efficient.

Epistemic processing involving experts

Resolving complex risk problems requires dialogue and deliberation among experts. Involving members of various epistemic communities which demonstrate expertise and competence is the most promising step for producing more reliable and valid judgements about the complex nature of a given risk. Epistemic discourse is the instrument for discussing the conclusiveness and validity of cause–effect chains relying on available probative facts, uncertain knowledge and experience that can be tested for empirical traceability and consistency. The objective of such a deliberation is to find the most cogent description and explanation of the phenomenological complexity in question as well as a clarification of dissenting views (e.g. by addressing the question of which environmental and socio-economic impacts are to be expected by specific actions or events). The deliberation among experts might generate a profile of the complexity of the given risk issue, based on selected subjectively chosen criteria. The deliberation may also reveal that there is more uncertainty and ambiguity than the initial appraisers had anticipated. It is advisable to include natural as well as social scientists in the epistemic discourse so that potential problems with risk perception can be anticipated. Controversies would thus be less as a surprise than they are now.

Reflective processing involving stakeholders

Characterising and evaluating risks, as well as developing and selecting appropriate management options for risk reduction and control in situations of high uncertainty, pose particular challenges. How can risk managers characterise and evaluate the severity of a risk problem when the potential damage and its probability are unknown or highly uncertain? Scientific input is therefore only the first step of a more sophisticated evaluation process. It is crucial to compile the relevant data and information on the different types of uncertainties, to inform the process of risk characterisation. The outcome of the risk characterisation provides the foundation for a broader deliberative arena, in which not only policy-makers and scientists but also directly affected stakeholders and public interest groups ought to be involved in order to discuss and ponder the 'right' balances and trade-offs between over- and under-protection. This reflective involvement of stakeholders and interest groups pursues the purpose of finding a consensus on the extra margin of safety that potential victims would be willing to tolerate and in which potential beneficiaries of the risk would be willing to invest, in order to avoid potentially critical and catastrophic consequences. If too much precaution is applied, innovations may be impeded or even eliminated; if too little precaution is applied, society may experience undesired consequences. The crucial question is how much uncertainty and ignorance the main stakeholders and public interest groups are willing to accept or tolerate in exchange for some potential benefit. The nature and scope of uncertainty may have implications for the range of groups to be involved and the procedures appropriate to address uncertainties and debate how decisions should be made. The reflective involvement of policy-makers, scientists, stakeholders and public interest groups can be accomplished through a spectrum of different procedures such as negotiated rule-making, mediation, round table or open forums, advisory committees and so on (Beierle and Cayford 2002; Klinke 2006; Rowe and Frewer 2000; Stoll-Kleemann and Welp 2006).

Participative processing involving the wider public

If risk problems are associated with high ambiguity, it is not enough to demonstrate that risk regulation addresses the issues of public concerns. In these cases, the process of evaluation needs to be open to public input and new forms of deliberation. This starts with revisiting the question of proper framing. Is the issue really a risk problem or is it an issue of life-style or future vision? Often the benefits are contested as well as the risks. The debate about designer babies illustrates the point that observers may be concerned not only about social risks of intervening in the genetic code of humans but also about the acceptability of the desired goal (Hudson 2006). Thus the controversy is often much broader than when dealing only with the direct risks. The aim is to find an overlapping consensus on the dimensions of ambiguity that need to be addressed in comparing risks and benefits, and balancing pros and cons. High ambiguity would require the most inclusive strategy for involvement because not only directly affected groups but also those indirectly affected should have an opportunity to contribute to this debate. Resolving ambiguities in risk debates necessitates involvement of the public to openly discuss competing arguments, beliefs and values. Participatory involvement offers opportunities to resolve conflicting expectations through a process of identifying overarching common values and defining options that allow a desirable life-style without compromising the vision of others. Critical to success is the establishment of equitable distribution rules for common resources and the activation of institutional means for achieving common welfare so that all can benefit. Possible procedures that involve the public include citizen panels or juries, citizen forums, consensus conferences, public advisory committees and similar approaches (Abels 2007; Beierle and Cayford 2002; Hagendijk and Irwin 2006; Klinke 2006; Renn 2008; Rowe and Frewer 2000).

An overview of the different participation and stakeholder involvement requirements with respect to linear, complex, uncertain and ambiguous risks is displayed in Figure 5.4 (in colour section). As with all classifications, this scheme shows a simplified picture of the involvement process and has been criticised as being too rigid in its linking of risk characteristics (complexity, uncertainty and ambiguity) and specific forms of discourse and dialogue (van Asselt 2005). In addition to the generic distinctions it may, for instance, be wise to distinguish between different types of risks and different types of regulatory cultures or styles (Löfstedt and Vogel 2001; Renn 2008). However, the purpose of this scheme is to provide a general orientation and to explain a generic distinction between ideal cases rather than to offer a strict recipe for participation.

The classification in Figure 5.4 offers a taxonomy of requirements for stakeholder and public inclusion based on the characteristics of risk knowledge. These general guidelines can be further specified by looking into each phase of the risk governance cycle (Renn and Walker 2008; Renn 2008).

Conclusion

This chapter has aimed to illustrate how the different components of pre-estimation, interdisciplinary risk estimation, risk characterisation, risk evaluation and risk management as well as communication and involvement interact with each other, and to demonstrate how the various combinations of complexity, uncertainty and ambiguity can be addressed by different risk management strategies.

The proposed modifications to the IRGC framework were based partly on critical remarks in the literature and partly on the experiences and lessons drawn from empirical case studies. The modified framework offers substantial advantages over conventional models of risk governance. First, it joins the two crucial elements of risk governance – analysis (i.e. knowledge generation) and management (i.e. decision-making). These two elements are kept separate from an analytical perspective but work hand in hand operationally. Second, the analytic distinction of risk characteristics – complexity, uncertainty and ambiguity – helps to tackle governance issues where complex technologies are concerned. This distinction not only highlights deficits in our knowledge about a risk issue but also shows the way forward by pointing out management options. Third, the risk governance framework includes public and stakeholder participation as well as risk communication as important functions in the risk governance process. The framework suggests efficient and adequate public/stakeholder participation. Concerns of stakeholders and/or the public are integrated in the risk appraisal phase via concern assessment. They are an established part of risk management. The optimum participation method thereby depends on the characteristics of the risk issue. Finally, the framework helps to identify governance gaps and deficiencies. At the same time, it proposes alternative actions.

The risk governance framework broadens stakeholders' views; it is an interactive, dialogical and systematic tool for reflecting the best available knowledge of physical risks as well as societal concerns, and for designing decision-making processes that consider the likely consequences of the management measures as well as their impact on the various groups in society.

References

Abels G (2007) Citizen involvement in public policy-making: does it improve democratic legitimacy and accountability? The case of pTA. *Interdisciplinary Information Science* **13**, 103–116.

Aven T and Renn O (2009) The role of quantitative risk assessments for characterising risk and uncertainty and delineating appropriate risk management options, with special emphasis on terrorism. *Risk Analysis* **29**, 587–600.

Aven T and Vinnem JE (2007) *Risk Management: With Applications from the Offshore Petroleum Industry.* Springer, Berlin.

Beierle TC and Cayford J (2002) *Democracy in Practice: Public Participation in Environmental Decisions.* Resources for the Future, Washington.

Bender HF(2008) Ergebnisse der Projektgruppe Risikoakzeptanz des AGS. *Gefahrstoffe-Reinhaltung der Luft* **68**, 287–288.

Brooks H (1984) The resolution of technically intensive public policy disputes. *Science, Technology, and Human Values* **9**, 39–50.

Daft RL and Weick KE (1984) Toward a model of organisations as interpretation systems. *Academy of Management Review* **9**, 284–295.

Feldman MS (1989) *Order without Design: Information Production and Policy Making.* Stanford University Press, Stanford.

Filar JA and Haurie A (Eds) (2010) *Uncertainty and Environmental Decision Making.* Springer, New York.

Frewer LJ and Salter B (2007) Societal trust in risk analysis: implications for the interface of risk assessment and risk management. In *Trust in Cooperative Risk Management: Uncertainty in Scepticism in the Public Mind.* (Eds M Siegrist, TC Earle and H Gutscher) pp. 143–158. Earthscan Publications, London.

Functowicz SO and Ravetz JR (1992) Three types of risk assessment and the emergence of post-normal science. In *Social Theories of Risk.* (Eds S Krimsky and D Golding) pp. 251–273. Praeger, London.

Goldstein J and Keohane RO (1993) Ideas and foreign policy: an analytical framework. In *Ideas and Foreign Policy: Beliefs, Institutions, and Political Change.* (Eds J Goldstein and RO Keohane) pp. 3–30. Cornell University Press, Ithaca.

Hagendijk R and Irwin A (2006) Public deliberation and governance: engaging with science and technology in contemporary Europe. *Minerva* **44**, 167–184.

Health Council of the Netherlands (2006) *Health Significance of Nanotechnologies.* Publication No. 2006/06E. Health Council of the Netherlands, The Hague.

HM Treasury (2005) *Managing Risks to the Public: Appraisal Guidance.* London. <www.hm-treasury.gov.uk/media/0/B/Managing_risks_to_the_public.pdf> (accessed May 2011).

Horlick-Jones T (2007) On the signature of new technologies: materiality, sociality, and practical reasoning. In *Risk and the Public Acceptance of New Technologies.* (Eds R Flynn and P Bellaby) pp. 41–65. Palgrave, London.

Horlick-Jones T and Sime J (2004) Living on the border: knowledge, risk and transdiciplinarity. *Futures* **36**, 441–456.

Horlick-Jones T, Rowe G and Walls J (2007) Citizen engagement processes as information systems: the role of knowledge and the concept of translation quality. *Public Understanding of Science* **16**, 259–278.

HSE (Health and Safety Executive) (2001) *Reducing Risk – Protecting People*. Health and Safety Executive, London.

Hudson KL (2006) Preimplantation diagnosis: public policy and public attitudes. *Fertility and Sterility* **58**, 1638–1645.

Hutter BM (2006) Risk, regulation, and management. In *Risk in Social Science*. (Eds P Taylor-Gooby and J Zinn) pp. 202–227. Oxford University Press, Oxford.

IRGC (International Risk Governance Council) (2005) *Risk Governance: Towards an Integrative Approach*. White Paper No. 1. International Risk Governance Council, Geneva.

Kahneman D and Tversky A (Eds) (2000) *Choices, Values, and Frames*. Cambridge University Press, Cambridge.

Keeney R and McDaniels T (2001) A framework to guide thinking and analysis regarding climate change policies. *Risk Analysis* **21**, 989–1000.

Klinke A (2006) *Demokratisches Regieren jenseits des Staates. Deliberative Politik im nordamerikanischen Große Seen-Regime*. Barbara Budrich Publishers, Opladen.

Klinke A and Renn O (2001) Precautionary principle and discursive strategies: classifying and managing risks. *Journal of Risk Research* **4**, 159–173.

Klinke A and Renn O (2002) A new approach to risk evaluation and management: risk-based, precaution-based, and discourse-based strategies. *Risk Analysis* **22**, 1071–1094.

Klinke A and Renn O (2010) Risk governance: contemporary and future challenges. In *Regulating Chemical Risks: European and Global Perspectives*. (Eds J Eriksson, M Gilek and C Ruden) pp. 9–27. Springer, Berlin.

Klinke A and Renn O (in press) Adaptive and integrative governance on risk and uncertainty. *Journal of Risk Research*.

Klinke A, Dreyer M, Renn O, Stirling A and van Zwanenberg P (2006) Precautionary risk regulation in European governance. *Journal of Risk Research* **9**, 373–392.

Laudan L (1996) The pseudo-science of science? The demise of the demarcation problem. In *Beyond Positivism and Relativism: Theory, Method and Evidence*. (Ed L Laudan) pp. 166–192. Westview Press, Boulder.

Lewin R (1992) *Complexity: Life at the Edge of Chaos*. Macmillan, New York.

Löfstedt RE (1997) *Risk Evaluation in the United Kingdom: Legal Requirements, Conceptual Foundations, and Practical Experiences with Special Emphasis on Energy Systems*. Working Paper No. 92. Center of Technology Assessment, Stuttgart.

Löfstedt R and Vogel D (2001) The changing character of regulation: a comparison of Europe and the United States. *Risk Analysis* **21**, 393–402.

Luhmann N (1993) *Risk: A Sociological Theory*. de Gruyter, Berlin.

Marti K, Ermoliev Y and Makowski M (Eds) (2010): *Coping with Uncertainty: Robust Solutions*. Springer, Berlin.

Nelkin D and Pollak M (1979) Public participation in the technological decisions: reality or grand illusion? *Technology Review* **6**, 55–64.

Nelkin D and Pollak M (1980) Problems and procedures in the regulation of technological risk. In *Making Bureaucracies Work*. (Eds CH Weiss and AF Burton) pp. 233–253. Sage, Los Angeles.

OECD (Organization for Economic Cooperation and Development) (2003) *Emerging Systemic Risks: Final Report to the OECD Futures Project.* OECD Press, Paris.

Pelling M, High C, Dearing J and Smith D (2008) Shadow spaces for social learning: a relational understanding of adaptive capacity to climate change within organisations. *Environment and Planning A* **40**, 867–884.

Radandt S, Rantanen J and Renn O (2008) Governance of occupational safety and health and environmental risks. In *Risks in Modern Society.* (Ed H-J Bischoff) pp. 127–258. Springer, Berlin.

Reese SD, Gandy OH Jr and Grant AE (Eds) (2003) *Framing Public Life: Perspectives on Media and Our Understanding of the Social World.* Lawrence Erlbaum, New Jersey.

Renn O (2008) *Risk Governance: Coping with Uncertainty in a Complex World.* Earthscan Publications, London.

Renn O and Schweizer P (2009) Inclusive risk governance: concepts and application to environmental policy making. *Environmental Policy and Governance* **19**, 174–185.

Renn O and Walker K (2008) Lessons learned: a re-assessment of the IRGC framework on risk governance. In *The IRGC Risk Governance Framework: Concepts and Practice.* (Eds O Renn and K Walker) pp. 331–367. Springer, New York.

Renn O, Klinke A and van Asselt M (2011) Coping with complexity, uncertainty and ambiguity in risk governance: a synthesis. *AMBIO* **40**, 231–246.

Rosa E (1997) Metatheoretical foundations for post-normal risk. *Journal of Risk Research* **1**, 15–44.

Rouse J (2011) The Chairman of the Joint Chiefs of Staff risk assessment system. Incorporation of the International Risk Governance Council framework. Paper presented at the Annual Meeting of the Society for Risk Analysis. Salt Lake City, Utah. 6 December. Arete Associates, Washington.

Rowe G and Frewer LJ (2000) Public participation methods: a framework for evaluation. *Science, Technology and Human Values* **25**, 3–29.

Stern PC and Fineberg V (1996) *Understanding Risk: Informing Decisions in a Democratic Society.* National Research Council, Committee on Risk Characterisation. National Academy Press, Washington.

Stirling A (2003) Risk, uncertainty and precaution: some instrumental implications from the social sciences. In *Negotiating Change.* (Eds F Berkhout, M Leach and I Scoones) pp. 33–76. Edward Elgar, London.

Stoll-Kleemann S and Welp M (Eds) (2006) *Stakeholder Dialogues in Natural Resources Management: Theory and Practice.* Springer, Berlin.

UK Treasury Dept (2005) Managing risks to the public: appraisal guidance. Draft for consultation. <www.hm-treasury.gov.uk> (accessed 11.02.11).

Underdal A (2009) Complexity and challenges of long-term environmental governance. *Global Environmental Change* **20**, 386–393.

US EPA (Environmental Protection Agency) (2009) *Potential Nano-enabled Environmental Applications for Radionuclides.* EPA-402-R-06-002. Environmental Protection Agency, Washington.

van Asselt MBA (2000) *Perspectives on Uncertainty and Risk.* Kluwer, Dordrecht.

van Asselt MBA (2005) The complex significance of uncertainty in a risk area. *International Journal of Risk Assessment and Management* **5**, 125–158.

Voss E and Wendler F (2009) Legal and institutional aspects of the general framework. In *Food Safety Governance: Integrating Science, Precaution and Public Involvement.* (Eds M Dreyer and O Renn) pp. 83–109. Springer, Berlin.

WBGU (German Advisory Council on Global Change) (2000) *World in Transition: Strategies for Managing Global Environmental Risks.* Springer, New York.

Wilkinson CF and Lamb JC (1999) The potential health effects of phthalate esters in children's toys: a review and risk assessment. *Regulatory Toxicology and Pharmacology* **30**, 140–155.

Zahariadis N (2003) *Ambiguity and Choice in Public Policy: Political Decision Making in Modern Democracies.* Georgetown University Press, Washington.

Part II

Dealing with risk at local scale

6 MARKET-BASED RESOURCE MANAGEMENT POLICY AND ENVIRONMENTAL UNCERTAINTY: OUTSOURCING RISK CALCULATION

Stewart Lockie

Markets and market-based instruments (MBIs) are promoted through contemporary policy and academic discourses as solutions to all manner of social and environmental problems. From improving health and aged care services to mitigating biodiversity loss and climate change, market-based approaches are presented as the most efficient and effective of available policy options. Thus we find tradable permits to emit greenhouse gases identified in virtually all key reports and international agreements on climate change mitigation as the instrument of choice (Garnaut 2008). Tax-based measures are seen to have some merit while regulatory interventions, incentive schemes and programs of direct action evoke notes of caution. Echoing the UK's Stern report, the Garnaut Climate Change Review commissioned by the Australian government to estimate the impacts of climate change and recommend policy frameworks to deal with it argued that non-market measures will be less efficient, impose unnecessary costs, distort investment decisions and result in no additional reduction in emissions across the economy (Garnaut 2008). The task, it argued, is to fix market failures, not pick winners.

At the time of writing, the politics of climate change mitigation suggest that adoption of greenhouse gas emissions trading in Australia is some way off. Resistance is similarly high in the US and several other countries with high and/or rising emissions. Despite this, market-based approaches have been trialled in numerous other arenas of environmental policy. This chapter will particularly focus on the implementation of MBIs within Australian agri-environmental policy and the assumptions and expectations these signify in relation to farmers' capacity to calculate and manage risk.

Agriculture is an inherently risky business. For the most part, farmers and graziers are price-takers with little control over the booms and busts of commodity price cycles. Couple this with terms of trade that have been in steady decline since the 1950s, and Australian farmers are left needing to boost their productivity (often through capital-intensive production technologies) to stay ahead of the cost-price squeeze and to manage their exposure to debt-related business risk. Climatic conditions compound these challenges by presenting farmers with a high degree of temporal and spatial variability in the short to medium term and with the prospect of more pronounced climatic shifts in the longer term. Variability in agricultural markets and in prevailing climatic conditions is entirely predictable. However, the timing, direction and magnitude of variability is characterised by a high degree of uncertainty. Farmers have historically adopted a range of strategies to deal with variability and uncertainty. Hay, silage and grain are stored. Pastures are destocked when dry weather sets in. Soil and water conservation works limit damage to natural resources. Mixed enterprises help deal with the ups and downs of commodity markets. Research and development (co-financed by farmers in Australia through industry levies) is often focused on technology to predict and cope with variability.

As many farmers argue, why wouldn't you protect your natural resources when they are the basis of your livelihood? Nevertheless, environmental degradation associated with agriculture often results from a failure to manage variability, i.e. to implement management strategies that are flexible and resilient enough to cope with the uncertain magnitude and timing of climatic and other events. An example is maintaining enough ground cover and/or constructing sufficient structural works to avoid soil erosion during all-too-infrequent heavy rainfall. Avoiding damage to farms' soil and water resources is not the only potential objective of agri-environmental policy. Reversing the degradation of endemic biodiversity has been a major, and highly controversial, focus of agri-environmental policy over the last decade with vegetation management plans and legislation introduced in most Australian states (Lockie 2009). While landscape biodiversity very likely provides ecosystem services of economic benefit to farmers (Lockie and Carpenter 2010), these are not well understood. The focus of vegetation management legislation has instead been on the protection of relatively natural ecosystems for their intrinsic and iconic value (Lockie 2010). The search for alternatives to regulatory restrictions on vegetation removal has seen biodiversity management emerge as a particularly popular policy objective for the application of MBIs.

MBIs, this chapter will argue, must be seen as one collection of technical policy instruments among many that have been deployed within a broader program of reform designed to expose farmers to market discipline and to develop their capacities as entrepreneurs and risk managers. On the surface, improving the ability of land managers to calculate and respond to environmental threats seems undeniably positive. This chapter will argue, however, that managing threats to environmental values and ecosystem processes is not necessarily a straightforward technical matter but one that may be characterised by considerable scientific, moral and political uncertainty.

Conceptualising and responding to risk and uncertainty

The opening chapter of this book provides an overview of the multiple ways in which risk and uncertainty may be conceptualised. This chapter will not attempt to arbitrate over whether definitive conceptualisations are desirable or possible. It will broadly follow the distinction made by Matthews (2009; see also Renn 2008a) and others between 'risk' as a possibility of harm for which a probability of occurrence may be estimated and 'uncertainty' as a possibility for which no probability may be estimated. The challenges thus evoked by risk and uncertainty for policy and management are quite different. Collective choice mechanisms such as markets and the private sector innovation they support generally deal well with quantifiable risks – actuarial analysis, insurance, hedging and numerous other strategies are used by businesses to calculate and manage risks (Matthews 2009). Markets deal less well, however, with uncertainty. Uncertainty, along with risks that are possibly low probability but catastrophic in consequence, necessitate both a heightened degree of caution and greater levels of institutional involvement in research and innovation (Matthews 2009). It is entirely appropriate, therefore, for governments to invest in basic research to identify future threats and to turn uncertainties into quantifiable risks in order to secure socially optimal outcomes. We might add, extending this argument, that it is equally appropriate for resource user groups, state agencies and other institutions to monitor compliance with resource use agreements, conventions and laws.

It is important to acknowledge that the line between risk and uncertainty may not always be clear. As Renn (2008b) pointed out, all calculations of risk embody some level of uncertainty. He argued that knowledge of risks is characterised by varying degrees of complexity (the relative level of difficulty involved in identifying and quantifying the cause and effect relationships between a risk event and its projected consequences), confidence[1] (the expected reliability and validity of predictions regarding those relationships) and ambiguity (controversy over what a risk means for those affected and the values that ought to applied in evaluating and managing the risk). Risk, in other words, is an inter-subjective and political concept just as surely as it is a scientific and technical concept. Turning uncertainties into quantifiable risks is thus not a solely technical exercise but one that simultaneously requires processes to understand perceptions of risk among different stakeholder groups and to facilitate deliberation and negotiation over the objectives of risk management (Renn 2009).

Uncertainty over the timing and magnitude of otherwise entirely predictable variability is, as the introduction to this chapter noted, a key feature of agriculture and the particular environmental threats faced by land managers. Risk theory suggests that shifting these threats from the domain of uncertainty to the domain of calculable risk ought to be one of the primary goals of research, capacity-building and policy innovation. Before examining particular attempts to build the calculative capacities of Australian farmers in relation to environmental threats, some additional points bear noting. First, agricultural production involves, by necessity, intervention in ecosystem processes and the potential to transmit negative (and positive) impacts across property and institutional boundaries as air, water, nutrients, sediments and organisms move through the landscape (Reeve 1997). The individual farm, administrative area or jurisdiction is never the single appropriate unit for the management of natural resources. Environmental threats associated with agriculture extend across the boundaries of private property and necessarily evoke the inter-subjective and political dimensions of risk calculation referred to above.

Second, scale effects here are critical. Whether the focus is on ecosystem processes and services or the meanings and values that are attached to them, very different priorities and goals potentially emerge at different spatial and temporal scales.

Agri-environmental policy and the market

Agri-environmental policy in Australia historically has largely been restricted to voluntary state- and territory-run programs of technical assistance. By the late 1980s, increasingly compelling evidence of the economic and environmental cost of land and water degradation created pressure for more proactive and national responses (Lockie 1999). At the same time, however, governments were engaged in an aggressive program of neo-liberal economic reform which saw, on the macro-economic front, the removal of import duties and export subsidies and, on the micro-economic front, the dismantling of statutory marketing boards, production quotas and other means of collectivising economic risk among farmers. The resolution to this apparent contradiction came through governmental innovations characterised as hybrid forms of neo-liberal governance in which social and environmental objectives were pursued through the parallel pursuit of economic rationality (Higgins and Lockie 2002). These programs adapted and changed over time in response to emergent contradictions and limitations (Lockie and Higgins 2007), resulting in at least three phases of Commonwealth involvement in agri-environmental governance. Drawing loosely on Hajkowicz (2009), these may be described as local capacity-building, building regional institutional capacity and experimenting with direct payments.

Phase 1 of Commonwealth involvement was exemplified by the National Landcare Program (NLP). Initiated in 1989, the NLP sought to promote and support the development of community Landcare groups and made limited funding available to assist in group coordination, establish experimental and demonstration sites, and provide training in property planning and other management techniques (Lockie 1999). The main focus of financial support, therefore, was on educational, research and planning activities that were believed likely to promote change, cooperation and investment among the wider farming community. Very little was made available for direct expenditure on environmental works. At the same time, the capacity-building features of the NLP were integrated with ostensibly economic programs. Most notable was the Rural Adjustment Scheme, a program designed principally to assist financially marginal farmers to exit the industry. In 1992, a new National Drought Policy was announced that challenged the view that drought was a 'natural disaster'. Instead, it defined climatic variability and dryness as predictable and manageable features of the Australian landscape, business risks that prudent and entrepreneurial farmers should be able to plan for and around (Higgins 2001). It followed that income support payments (i.e. welfare measures) should be replaced with capacity-building programs for those farmers considered financially viable or likely to achieve viability, and with structural adjustment programs for those who were not. Direct subsidies for participation in property and catchment planning were subsequently made available through the Rural Adjustment Scheme to farmers applying through community Landcare groups. Although the complete removal of payments to farmers in what became known as 'exceptional circumstances' proved politically difficult, the linking of the NLP, National Drought Policy and Rural Adjustment Scheme reshaped the circumstances in which risk was considered to be individual rather than social. Further, it redefined circumstances where risk was still determined to be social as opportunities to encourage farmers in 'temporary difficulties' to develop their 'future capacities for profitability' (Higgins 2001, p. 312).

At face value, the NLP looks very much like 'regulatory dumping'(Peck and Tickell 2002) – devolutionist approaches to governance through which local governments or communities are given responsibility to deal with social and/or environmental problems without the requisite power or resources to fulfil that responsibility. However, community Landcare groups were not seen simply as a low-cost mechanism through which to diffuse information and promote attitude change. The networking and capacity-building activities of these groups were seen to encourage social learning and risk-taking that capitalised on and respected the local knowledge of Landcare group members, encouraged landholders to coordinate their activities on a catchment or subcatchment basis, made better use of the resources of state government agencies and the agribusiness sector and, ultimately, reduced the personal and financial costs to landholders of dealing with degradation (Lockie 1998; Scott 1998). The NLP also largely avoided conflict with farmers over the extent of their perceived private property rights (Reeve 2001) and the difficulties inherent in regulating diffuse forms of resource use (Industry Commission 1997).

The NLP was enormously successful in enrolling landholders (some 30% of Australian farm businesses) and in encouraging farm-scale improvements in natural resource management (ABARE 2003). However, the ecological and economic complexities of natural resource management led to slow progress in turning the consensus and excitement over Landcare into landscape-scale environmental improvements (CSIRO 2003). Meanwhile, Australia's competitors used public support for environmental protection to argue that, by shifting subsidies away from the production of agricultural commodities and towards the protection of other cultural and environmental values on behalf of the community, these subsidies no longer distorted trade (Lockie and Goodman 2006). In concert with protracted, and perhaps climate change-induced, drought conditions over much of Australia's key agricultural regions, these challenges rendered the political solution exemplified by Landcare to the environmental contradictions of productivist agriculture increasingly problematic.

Phase 2 of Commonwealth involvement in agri-environmental policy sought to address concerns that tight terms of trade made it difficult for landholders to undertake works where the benefits were primarily off-site, long-term or unlikely to boost productivity (Lockie 2006). In 1999, the Commonwealth proposed to devolve greater levels of funding to the regional level for natural resource planning and investment and to increase incentives for improved management by individual landholders through the use of MBIs (Dibden and Cocklin 2005; Hajkowicz 2009). Fifty regional natural resource management bodies were subsequently recognised or established and –reflecting the hybrid approach to governance pioneered in the NLP –charged with enabling the community to decide for itself the appropriate mix of regulatory instruments and MBIs. At the same time, they were subjected to heightened levels of monitoring and reporting on expenditure and changes in resource condition. In reality, use of regulatory instruments was the domain of state and territory governments while regional groups focused on voluntary programs. Uptake of MBIs as delivery mechanisms for those programs was slow. A national MBI pilot program funded 11 projects between 2003 and 2005 to encourage testing and development by regional groups (NMBIWG 2005). Four pilot projects used auction systems to direct payments for biodiversity conservation and five projects used cap-and-trade and/or offset schemes to address soil salinity and water quality. Other projects examined conservation insurance and leveraged investments. Nine Round 2 projects announced in 2006 focused on the refinement of auction and offset instruments to improve cost-effectiveness, increase participation, deal better with uncertainty, ensure compliance and so on.

Phase 3 of Commonwealth involvement followed the announcement in 2007 of a national Environmental Stewardship Program (Hajkowicz 2009), under which landholders were to 'be paid to undertake agreed actions beyond their regulated responsibilities to achieve public benefit environmental outcomes' (Australian Government 2007). Environmental services purchased through the Stewardship Program were to include the protection of nationally endangered ecological communities and natural values associated with world and national heritage places. The program has been implemented very carefully to ensure that direct payments are made only for the provision of values considered unambiguously public in nature; in practice, the preservation and/or restoration of box gum grassy woodland ecosystems through agreed actions including reduced stocking and grazing intensities, reduced fertiliser use, weed management and native species planting (Australian Government 2010). The intrinsic value of these ecosystems is not in dispute here. The important point is that the Phase 3 focus on providing incentive payments to protect environmental values such as endemic biodiversity from agriculture – as opposed to the development of more biologically diverse agro-ecosystems that support agriculture – marks a distinct break from the hybrid approach to agri-environmental governance exemplified by the NLP which promoted environmentally responsible and productive agriculture as more or less the same thing. While the use of market-like allocation mechanisms has given implementation of the National Stewardship Program an air of economic rationality, the program has actually been targeted at environmental values deemed unlikely to be protected as the logical outcome of economically rational natural resource utilisation. In part, this has been to ensure that payments are not made for environmental services that support agricultural production and thereby provide private benefits which would be inconsistent with the neo-liberal project and its associated free trade agenda. Somewhat paradoxically, stewardship payments could, if scaled up, potentially create a significant alternative income stream that might assist landholders to manage financial and other business risks.

The logic of market failure

Explanations for natural resource degradation in predominantly agricultural landscapes have become increasingly dominated in Australia by the argument that degradation arises as a consequence of market failure. According to this logic, soils, water, ecosystem processes and so on provide the essential conditions of production for agriculture. The costs of conserving these resources are therefore costs of production that, in a properly functioning market, will be internalised by producers and passed on to consumers (Commonwealth Treasury, n.d.). Economists, however, identify three broad sources of market failure that impede this internalisation and transfer of costs (Scott 1998). First, imperfect or asymmetrical information results in inadequate understanding of the long-term impact of resource use practices among users and/or regulators. Second, inadequately defined property rights remove individuals' incentive to conserve, since costs are borne individually while benefits are distributed among all resource users. Third, pricing of natural resource inputs below their full economic and environmental cost leads to overutilisation. It follows that the optimal solution to environmental degradation under market-based production is always to address these sources of market failure. It is never to impose regulatory restrictions on resource use or to subsidise improved management practices; the former is decried as blunt and inflexible, the latter as creating potential de facto barriers to trade that deny landholders' duty of care not to degrade resources or cause harm to others (Productivity Commission 2004).

Of course, it is not always possible to fix markets. Even where resource degradation is well understood, clearly attributable and imposes direct costs on producers there is no guarantee that producers will be able to pass on to consumers any costs they do internalise. Eco-labelling schemes are one mechanism through which this may be achieved, but these currently account for only a small share of agricultural production and retail sales. The vast majority of producers retain a marginal position as price-takers within highly competitive mass commodity markets. In the case of public goods, including extremely long-term or intergenerational environmental benefits, markets simply do not exist (Bromley 1989). With no market, there is nothing to fail. While it is reasonable to expect landholders to provide a certain amount of public good conservation as a condition of resource access, the socially optimal level of conservation may well exceed what might be deemed industry best practice and/or the financial and managerial capacities of individual landholders.

In such circumstances, the establishment of markets for the provision of ecosystem services is preferred over regulation or direct subsidies in order to ensure that public and private expenditures are directed to wherever the relevant service can be provided at least cost. MBIs include a mixture of trading mechanisms, auctions and price signals designed to influence people's behaviour in pursuit of policy objectives (Scott 1998). Mechanisms are generally classified into three major groups: market friction mechanisms such as eco-labelling or education that seek to improve the efficiency of existing markets by removing obstacles to recognition of ecosystem services, price-based mechanisms such as auctions, tenders and taxes that set or modify prices to incorporate the cost of ecosystem services, and quantity-based mechanisms such as cap-and-trade and offset schemes which set targets to achieve or maintain environmental services (Whitten and Shelton 2005; NMBIWG 2005). Auctions and tenders are believed particularly suitable for diffuse source environmental outcomes such as terrestrial biodiversity.

According to their proponents, MBIs offer the technical means with which to design solutions to 'previously intractable policy problems' (NMBIWG 2005, p. 5). They achieve this, it is argued, by allowing flexibility for individuals to choose the optimum amount and means of conservation depending on their circumstances, the least-cost path to overall environmental outcomes, and the provision of continuing incentives to find innovative ways to further reduce environmental impacts (Commonwealth Treasury, n.d.; Productivity Commission 2004). MBIs therefore rely on the same calculative capacities among landholders that Phase 1 agri-environmental programs such as Landcare and the National Property Planning Program sought to develop. To calculate the worth of target environmental values to their own businesses and the incentive payment necessary to provide additional conservation, landholders need to understand the key natural resource components under their management (e.g. soils, waterways, vegetation communities), the contribution of these components to ecosystem functioning and agricultural production, the wider social values associated with these components and functions (e.g. provision of clean water, protection of endangered species) and their own personal and business goals. The Environmental Stewardship Program and several of the MBI pilot programs have thus included a range of advisory and capacity-building elements to support landholder participation. Evaluations of these programs suggest that increased contact with natural resource management agency staff, along with resource assessment and planning skills, are among the most highly valued outcomes for participants (Lockie and Tennent 2010).

MBIs and risk calculation

MBIs, as exemplified by the Environmental Stewardship Program and several MBI pilot programs, offer financial incentives for the public good component of endemic biodiversity conservation to landholders willing to provide conservation of suitable quality at least cost. To achieve least-cost conservation, these programs rely (whether or not they explicitly acknowledge it) on participating landholders' understanding that endemic biodiversity and relatively 'natural' ecosystem components also provide a variety of private goods. These range from landholders' personal interests and/or sense of responsibility to the provision of ecosystem processes and services that actually support agricultural production (Commonwealth Treasury, n.d.). A key concept is functional biodiversity, defined as the relationships between groups of organisms (e.g. bacteria), the ecosystem functions they perform (e.g. decomposition) and the ecosystem goods or services these functions provide (e.g. nutrient cycling) (Swift *et al.* 2004). It must also be noted that biodiversity at the micro-scale can be expected to vary significantly over time due to the combined effects of changing environmental conditions and the variable capacity of species to adapt, move and/or colonise new ecological niches (Swift *et al.* 2004; Zimmerer 1994). In the short term, therefore, absolute species decline at a given locale may not lead to a corresponding decline in ecosystem function provided major species groups and relationships are maintained, particularly at the ecosystem or landscape levels (Swift *et al.* 2004). Continued degradation, however, may induce threshold effects and the rapid failure of critical ecosystem processes and services. Absolute species diversity thus provides substitutability between species and ecosystem-level adaptability in the longer term.

While many agricultural technologies substitute ecosystem services provided by functional biodiversity with mechanical or chemical inputs (e.g. fertiliser) in order to boost the productivity of cultivated species, there is evidence that extensive and intensive agricultural systems alike benefit from management designed to increase non-cultivated species diversity (Dorrough *et al.* 2004; Omer *et al.* 2010). Degradation of biodiversity can lead eventually to increased input costs and reduced productivity. Given the climatic variability inherent to Australian environments, likely to increase as a consequence of anthropogenic climate change, biodiversity degradation can reasonably be expected to undermine ecosystem resilience and lead to increased climate change adaptation costs. It would appear that biodiversity conservation, in a general sense, is very much in landholders' collective self-interest. It would also appear that biodiversity management, alongside other aspects of natural resource management such as soil conservation, ought to be a key component of individual landholders' environmental and business risk management. This returns us to the question of landholders' capacity for risk calculation, in particular the roles of complexity, confidence and ambiguity in such calculation.

Complexity, as used by Renn (2008b), refers to the degree of difficulty involved in quantifying cause–effect relationships between a risk event and its projected consequences. The notion of biodiversity is itself a proxy for highly complex relationships between organisms and groups of organisms across space and time. The role of these relationships in supporting ecosystem function, as discussed above, led Swift *et al.* (2004) to hypothesise that ecosystem processes may be maintained, if not enhanced, following reductions in biodiversity at the farm and field scale provided these reductions are not mirrored at a higher scale. The importance of mosaics of land uses and habitat types at a landscape scale appears to make this the natural unit of management for biodiversity, although some research has demonstrated a positive impact on agricultural productivity within intensive cropping systems from field-scale biodiversity-friendly landscape components such as uncultivated field margins and hedgerows (Omer *et al.* 2010). Beyond general principles, therefore, much remains unknown about the contribution of endemic biodiversity and relatively natural ecosystem components to the productivity and resilience of agricultural systems and, as a consequence, the optimal degree to which these might be conserved or built back into agricultural landscapes at a variety of scales (Lockie and Carpenter 2010). Compounding this complexity is the spatial and temporal mismatch between ecosystem processes in which biodiversity is implicated, and the private property rights of landholders. While the costs of conservation are borne by individuals *in situ*, the benefits for agricultural productivity and ecosystem resilience are distributed, to varying degrees, across the wider landscape and across time as species move across habitats, colonise new niches and so on. This creates an incentive to treat biodiversity as an open access resource (i.e. a resource subject to no legal or institutional framework governing access, management or exclusion) and thus to free-ride on the conservation efforts of others.

Confidence refers to the expected reliability and validity of predictions regarding relationships between risk events and consequences (Renn 2008b).[1] As above, the general proposition that biodiversity at a number of scales will be critical to the long-term sustainability and productivity of agricultural systems in a future marked by even more variability than has been experienced historically, appears sound. This, however, is like observing that the relationships between atmospheric concentrations of greenhouse gases, temperatures and rainfall are well understood. They are. But this understanding does not, by itself, resolve uncertainty over the magnitude and timing of changes in the global climate system nor, critically, the magnitude and timing of more localised systemic changes or of specific climate events. Application of the precautionary principle would suggest that landholders ought to plan for adverse events, plan more biodiversity into their farming systems than probably necessary for adaptation to a changing climate, and collaborate with their neighbours to scale these planning efforts up to a more appropriate scale. These may be useful rules of thumb, but they also suggest that what we are talking about is the management of uncertainty as opposed to the management of calculable risk.

Ambiguity refers to controversy over what a risk means for those affected and the values that ought to be applied in evaluating and managing the risk (Renn 2008b). Access to property or other natural resources presupposes a bundle of obligations as well as a bundle of rights (Reeve 1997). Generally, these obligations include prohibitions on uses of the resource that are harmful to others or that otherwise degrade it. Enactment of regulatory restrictions on tree clearing in several Australian states reflect the view among state governments that degradation of biodiversity defies landholders' duty of care. Conversely, the implementation of MBI programs to provide incentives for public good biodiversity outcomes suggests limits to this duty. There is no clear resolution regarding questions of how much responsibility private landholders ought to bear for providing ecosystem services to future generations, the wider community or even their immediate neighbours.

While this chapter will not examine empirical evaluations of existing MBI schemes in detail, there is some evidence to suggest that the arguments presented above have been reflected in the experiences of MBI project participants. Official evaluations of MBI projects have stressed their success in securing agreements with landholders to protect target ecosystems at a cost substantially lower than some alternatives (e.g. the establishment of reserves and national parks) (Grafton 2005). However, independent studies have raised questions about the ability of these projects to establish genuinely competitive bidding processes, provide sufficient financial incentive to enrol landholders who not have provided the required service anyway, guarantee long-term outcomes or pass the costs of protecting natural resources along the value chain (Lockie and Tennent 2010). Lockie and Tennent's (2010) study also raised questions about the ability of MBI schemes to target multiple environmental objectives, a requirement that depends on the potential ability of multiple landholders to provide the desired service. The particular MBI pilot project they examined targeted two objectives – protection of specific plant communities and connectivity between protected areas across the bioregion. To secure the second objective, compromises were made to the first. Cocklin *et al.* (2006) reported that Victorian farmers participating in similar projects thought it difficult to untangle the public and private benefits of environmental care and to distinguish between what should be included in the duty of care inherent in private property rights and what should reasonably be subsidised as additional public goods.

Conclusion

MBIs have a great deal to offer as one group of options among many in resource management and agri-environmental policy where basic conditions for their effective and efficient operation can be met. Agencies and participants ought to have a sound and reliable understanding of cause–effect relationships between management actions and desired ecosystem services. There should be political and community consensus, at the appropriate scales, that public investment in these actions is justified. Actions by individual resource users ought to be capable of scaling up to landscape-level outcomes. At the same time, lack of conservation activity by non-participants should not undermine these outcomes. Actions and outcomes must be monitorable. Sufficient resource users ought to be capable of providing the required service to enable a genuine market to operate. Non-corroboration between resource users in the setting of market prices should not undermine their participation in collaborative planning exercises or, if this is unavoidable, non-corroboration should be demonstrably more efficient. Resource users should not, as a group, be facing short-term financial pressures so acute as to lead them to discount too heavily the future private benefits of ecosystem services.

The narrative of market failure and the superiority of MBIs in resolving intractable policy problems seldom acknowledges such conditions. Indeed, this narrative is elegant in its simplicity. Complex and potentially divisive issues are to be dealt with through the application of technical expertise (in the design of markets) and the entrepreneurial and calculative capacities of individual landholders. The flexibility and cost-effectiveness argued to characterise MBIs consequently presuppose landholders' capacity to calculate how the costs and benefits of biodiversity conservation or enhancement are distributed and how much uncompensated responsibility they ought to bear for the provision of public benefits. This chapter has argued that there is plenty of evidence to support, as a rule of thumb, the proposition that biodiversity and other natural resources ought to be conserved and/or designed into agricultural production landscapes at a variety of scales in order to manage both short-term environmental variability and longer-term climate risks. However, there is sufficient complexity, ambiguity and lack of confidence regarding the specific relationships between conservation actions and benefits at a variety of scales (spatial and temporal) to suggest that landholders are not actually in a position to calculate either the current value of ecosystem services provided by biodiversity or the future costs of postponing management action to enhance it. Landholders are not, on the whole, managing calculable risks. They are managing uncertainties.

As much as it may be in landholders' interest to build biodiversity management into their regular business risk management, the high level of uncertainty around cause–effect relationships between biodiversity management actions and environmental risks, the multiple spatial and temporal scales at which these relationships play out, and the ambiguity over where landholders' duty of care ends and public responsibility begins, make it clear that there remains a strong role for governments and the wider public in resolving uncertainty and in supporting management actions. Market-based environmental policy does not deny the role of public investment, and individual MBI projects are very specific about the sorts of environmental values that need additional protection in the public interest. However, calculating the appropriate incentive to support provision of this additional protection is still outsourced to individual landholders. The importance of involving private landholders in risk calculation and management is not in dispute (Renn 2009). In fact, such involvement is critical. The point is that outsourcing risk calculation to private landholders does not, and cannot, resolve uncertainty over fundamentally moral and political issues such as priorities for conservation on the basis of iconic, intrinsic, amenity, heritage or other social and cultural values, or the balance between rights and obligations that are inherent in private property. Nor is it appropriate to entirely individualise arrangements for the delivery of agri-environmental programs when natural resources and climate change adaptation require coordinated management at higher spatial and temporal scales.

Endnote

1 Renn (2008b) actually uses the term 'uncertainty' to refer to the reliability and validity of predicted cause–effect relationships. 'Confidence' is the preferred term in this chapter, because its everyday usage is a better fit with the definition provided in relation to risk and to avoid confusion in relation to the distinction drawn between risk and uncertainty.

References

ABARE (Australian Bureau of Agricultural and Resource Economics) (2003) *Natural Resources Management on Australian Farms*. Australian Bureau of Agricultural and Resource Economics, Canberra.

Australian Government (2007) *Environmental Stewardship Program Strategic Framework 2007*. Dept of Agriculture, Fisheries and Forestry and Dept of the Environment and Water Resources, Canberra.

Australian Government (2010) Environmental stewardship. <www.nrm.gov.au/stewardship/index.html> (accessed 28.12.10).

Bromley D (1989) Entitlements, missing markets, and environmental uncertainty. *Journal of Environmental Economics and Management* **17**, 181–194.

Bromley D (2007) Environmental regulations and the problem of sustainability: moving beyond 'market failure'. *Ecological Economics* **63**, 676–683.

Cocklin C, Dibden J and Mautner N (2006) From market to multifunctionality? Land stewardship in Australia. *Geographical Journal* **172**, 197–205.

Commonwealth Treasury (no date) *Public Good Conservation and the Impact of Environmental Measures Imposed on Landholders*. Commonwealth of Australia, Canberra.

CSIRO (Commonwealth Scientific and Industrial Research Organisation) (2003) *Assessing the Impact of Landcare Activities on Natural Resource Condition. Attachment 4. Review of the National Landcare Program*. Dept of Agriculture, Fisheries and Forestry, Canberra.

Dibden J and Cocklin C (2005) Agri-environmental governance. In *Agricultural Governance: Globalization and the New Politics of Regulation*. (Eds V Higgins and G Lawrence) pp. 135–152. Routledge, London.

Dorrough J, Yen A, Turner V, Clark G, Crosthwaite J and Hirth J (2004) Livestock grazing management and biodiversity conservation in Australian temperate grassy landscapes. *Australian Journal of Agricultural Research* **55**, 279–295.

Dumaresq D, Carpenter D and Lockie S (2010) The human ecology of agrobiodiversity. In *Agriculture, Biodiversity and Markets: Livelihoods and Agroecology in Comparative Perspective*. (Eds S Lockie and D Carpenter) pp. 33–45. Earthscan Publications, London.

Garnaut R (2008) *The Garnaut Climate Change Review: Final Report*. Cambridge University Press, Melbourne.

Grafton Q (2005) *Evaluation of Round One of the Market Based Instrument Pilot Program*. National Action Plan for Salinity and Water Quality, Canberra.

Hajkowicz S (2009) The evolution of Australia's natural resource management programs: towards improved targeting and evaluation of investments. *Land Use Policy* **26**, 471–478.

Higgins V (2001) Calculating climate: 'advanced liberalism' and the governing of risk in Australian drought policy. *Journal of Sociology* **37**, 299–316.

Higgins V and Lockie S (2002) Re-discovering the social: neo-liberalism and hybrid practices of governing in natural resource management. *Journal of Rural Studies* **18**, 419–428.

Industry Commission (1997) *A Full Repairing Lease: Inquiry into Ecologically Sustainable Land Management*. Industry Commission, Melbourne.

Lockie S (1998) Landcare in Australia: cultural transformation in the management of rural environments. *Culture and Agriculture* **20**, 21–29.

Lockie S (1999) The state, rural environments and globalisation: 'action at a distance' via the Australian Landcare Program. *Environment and Planning A* **31**, 597–611.

Lockie S (2006) Networks of agri-environmental action: temporality, spatiality and identity within agricultural environments. *Sociologia Ruralis* **46**, 22–39.

Lockie S (2009) Agricultural biodiversity and neoliberal regimes of agri-environmental governance in Australia. *Current Sociology* **57**, 407–426

Lockie S (2010) Neoliberal regimes of environmental governance: climate change, biodiversity and agriculture in Australia. In *The International Handbook of Environmental Sociology*. 2nd edn. (Eds M Redclift and G Woodgate) pp. 364–377. Edward Elgar, London.

Lockie S and Carpenter D (2010) Agrobiodiversity and sustainable farm livelihoods: policy implications and imperatives. In *Agriculture, Biodiversity and Markets: Livelihoods and Agroecology in Comparative Perspective*. (Eds S Lockie and D Carpenter) pp. 303–313. Earthscan Publications, London.

Lockie S and Goodman M (2006) Neoliberalism, standardisation and the problem of space: competing rationalities of governance in fair trade and mainstream agri-environmental networks. In *Between the Local and the Global: Confronting Complexity in the Contemporary Agri-Food Sector.* (Eds T Marsden and J Murdoch) pp. 95–117. Elsevier, Oxford.

Lockie S and Higgins V (2007) Roll-out neoliberalism and hybrid practices of regulation in Australian agri-environmental governance. *Journal of Rural Studies* **23**, 1–11.

Lockie S and Tennent R (2010) Market instruments and collective obligations for on-farm biodiversity conservation. In *Agriculture, Biodiversity and Markets: Livelihoods and Agroecology in Comparative Perspective.* (Eds S Lockie and D Carpenter) pp. 287–301. Earthscan Publications, London.

Matthews M (2009) Fostering creativity and innovation in cooperative federalism: the uncertainty and risk dimensions. In *Critical Reflections on Australian PublicPpolicy: Selected Essays.* (Ed J Wanna) pp. 59–70. ANU Press, Canberra.

NMBIWG (National Market-Based Instrument Working Group) (2005) National Market-Based Instruments Pilot Program: Round One. Interim report. National Action Plan for Salinity and Water Quality, Canberra.

Omer A, Pascual U and Russell N (2010) The contribution of biodiversity to modern intensive farming systems. In *Agriculture, Biodiversity and Markets: Livelihoods and Agroecology in Comparative Perspective.* (Eds S Lockie and D Carpenter) pp. 117–135. Earthscan Publications, London.

Peck J and Tickell A (2002) Neoliberalizing space. *Antipode* **34**, 380–404.

Productivity Commission (2004) *Impacts of Native Vegetation and Biodiversity Regulations.* Commonwealth of Australia, Canberra.

Reeve I (1997) Property and participation: an institutional analysis of rural resource management and Landcare in Australia. In *Critical Landcare.* (Eds S Lockie and F Vanclay) pp. 83–95. Centre for Rural Social Research, Charles Sturt University, Wagga Wagga.

Reeve I (2001) Property rights and natural resource management: tiptoeing round the slumbering dragon. In *Rurality Bites: The Social and Environmental Transformation of Rural Australia.* (Eds S Lockie and L Bourke) pp. 257–269. Pluto Press, Sydney.

Renn O (2008a) Concepts of risk: an interdisciplinary approach. Part 1: Disciplinary risk concepts. *GAIA* **17**, 50–66.

Renn O (2008b) Concepts of risk: an interdisciplinary approach. Part 2: Integrative approaches. *GAIA* **17**, 196–204.

Renn O (2009) Inclusive risk governance: concepts and application to environmental policy making. *Environmental Policy and Governance* **19**, 174–185.

Scott F (1998) Market-based approaches for sustainability. In *Sustainability in a Commercial Context: Market-based Approaches.* (Eds F Scott, G Kaine, R Stringer and K Anderson) pp. 33–59. Land and Water Resources Research and Development Corporation, Canberra.

Stern N (2007) *The Economics of Climate Change: The Stern Review.* Cambridge University Press, Cambridge.

Swift M, Izac A and van Noordwijk M (2004) Biodiversity and ecosystem services in agricultural landscapes: are we asking the right Questions? *Agriculture, Ecosystems and Environment* **104**, 113–134.

Whitten S and Shelton D (2005) *Markets for Ecosystem Services in Australia: Practical Design and Case Studies.* CSIRO, Canberra.

Zimmerer K (1994) Human geography and the 'new ecology': the prospect and promise of integration. *Annals of the American Association of Geographers* **84**, 108–125.

7 WOMEN AND RISK: COMMERCIAL WASTEWATER INJECTION WELLS AND GENDERED PERCEPTIONS OF RISK

Lee M. Miller

> So, I mean, when there's a question you should err on the side of safety, not on the side of 'maybe'!

In 1995, in a rural area of south-east Texas just outside the city limits of Conroe, a retired beautician and her husband learned from their nephew that a commercial wastewater injection well was being drilled on the land behind their home. They were concerned about the prospect of wastewater being injected under high pressure into the ground only feet from their private water well and only source of water.[1] Inquiries to the Conroe water department revealed that there was no record of an injection well despite the requirement that the water department be notified upon application for an injection well permit. This information made the couple more uneasy. He was facing health issues, making the news even more disturbing. She swung into action to rally their neighbours and family. A small group of residents formed to voice their objections to the wastewater injection well.

Although the wastewater injection well permit was issued and the well completed in the mid 1990s, the company went bankrupt and the well was never used. The property was abandoned with nothing to indicate the well except the well head and several small storage tanks, barely noticeable from the road. The neighbours were relieved. Approximately 10 years later, a second company purchased the property with the injection well and made plans to drill three additional injection wells on the same lot. Fearing possible water, soil and air contamination and associated risks to health, well-being and property values, the neighbourhood group formed again to oppose the wells.

Since 2005, residents surrounding the wastewater injection well site have been successful in garnering the support of Montgomery County officials, Conroe city officials, the area's state representative and senator, the regional groundwater conservation district and neighbourhood associations over an expanded geographic area. Although it took some time to achieve, there is now total consensus among all local stakeholders against the injection wells. This makes the case study somewhat unusual and explains the fact that, to date, the permitting process has been delayed for over six years and the company has not started operation of the existing well. Despite the local group's success at delaying the arrival of the wells, at the time of writing it looks likely that the commercial wastewater injection wells will be allowed to operate.

Several men have played prominent formal roles in local opposition to the wells, but more than two-thirds of the core organising/activist group members were women. They explained their involvement in terms of risk perceptions based on gender roles, personal experiences and conviction that uncertainty should trigger caution. With two exceptions, the men involved in the case against the wastewater injection wells occupied professional roles (legal, administrative, political, technical) that determined their involvement with the case. Perhaps it is due to these roles that their perspective on the risks of the wells was discussed in formalised ways using technical language. The male members of the residents' group opposing the wells also framed their ideas about risk in technical ways, very different from the ways women spoke of risk. The perceptions of risks associated with the planned arrival of commercial wastewater injection wells are the central themes of this chapter.

Wastewater injection well background

Texas is home to more injection wells than any other state, with 20% of all US injection wells (EPA 2006). In Texas, the use of injection wells developed simultaneously with the oil and gas industry in the 1930s. Saltwater used in production of oil and gas was disposed of by pumping it back into the formations from which the oil and gas were extracted. These injection wells are classified as Class II injection wells and account for the majority of injection wells in the state.[2] Class II injection wells are regulated by the Texas Oil & Gas Commission.[3] Class I injection wells employ the same technology to pump hazardous and non-hazardous liquid waste thousands of metres into the ground, well below underground sources of drinking water. According to federal law, elements injected by Class I wells are considered non-hazardous if they are not ignitable, corrosive, reactive or toxic at the point of disposal (EPA 2001).[4] The 'non-hazardous' classification is seen by the participants in this study as problematic because the chemicals are hazardous if sufficient dilution does not occur prior to injection. Residents worry that, since there are no external monitoring mechanisms in place to ensure the proper dilution takes place, the wastewaters injected may in fact be very dangerous.[5]

The study

Grangerland, Texas, is an unincorporated area immediately north-east of the city of Conroe in south-east Texas. Conroe is the county seat of Montgomery County, the northernmost county in the greater Houston metropolitan area. The southern and western portions of the county closest to Houston are more densely populated and wealthier than the rest of the county. The main north–south artery, Interstate 45, divides the county. Neighbourhoods to the west of the highway enjoy relatively more money and political clout, areas to the east are poorer and more rural. The wastewater injection well site lies to the east of I45, north of Conroe.

The case study began as a project to identify factors involved in community ability to mobilise against technological and environmental hazards. Since it was aimed at exploring possible aids and obstacles to community resistance, a case study approach was adopted (Berg 2007; Gunter and Kroll-Smith 2007; Yin 2009). As respondents talked about their reasons for opposing the wells, they voiced their views on the risks they felt the wastewater injection wells presented.

The specific case was chosen for several reasons. The first was that the community efforts to fight the wastewater injection wells came to me. One of my students lived in the affected area, was a supporter of the group's efforts and told me about the grassroots initiative to block the wells. I had some previous familiarity with the literature on conflicts about where to locate facilities seen as potentially hazardous to the environment (siting disputes) and was interested in studying community dynamics surrounding them. This group was in the middle of oppositional efforts, which would allow me to study the factors shaping collective resistance to the wastewater wells as they emerged.

Another important factor in the selection of this case was proximity. The community involved is located just south of the county in which I live so travel time to interviews, meetings, hearings and fundraising events averaged about 30 minutes by car. For background information on the local economy and underground injection wells, I consulted historical and environmental documents, technical documents on injection well technology and case histories of other communities facing similar issues. I also spoke with representatives from several other Texas communities with a history of disputes over wastewater injection wells.

Original data were gathered through semi-structured interviews with residents involved in the injection well opposition. The interviewees represented a variety of roles. Most, but not all, were employed. Several were entrepreneurs with small businesses. A handful worked in, or had retired from, the oil and gas industry. Roughly two-thirds (17) of those interviewed were women. Ages ranged from mid 30s to late 70s. All respondents were white, reflecting the racial make-up of the neighbourhood surrounding the well site and the membership of the grassroots group resisting its development.

People whose names appeared in news articles and legal documents associated with the community opposed to the injection wells were the initial contacts. Other contacts were suggested by each respondent in a snowball sample, leading to a total of 26 interviewees. Interviews included the respondents' perceptions of the threat from the wells. Each person was asked how they heard about the wells and how they felt about their arrival. They were also asked how they became involved in the opposition to the wells and their motivations for taking action. Information about each person's experience with environmental issues and activism was gathered. Respondents also provided documents, newspaper articles, case documentation and correspondence. Technical and environmental reports were consulted for knowledge about the specific case.

Interviews averaged about an hour, although some lasted over four hours. Each interview was digitally recorded and transcribed. Guided by grounded theory (Strauss and Corbin 1990), emerging themes were identified through the process of coding the interview transcripts and field notes. Initial coding noted basic information: personal information about respondents, case history of injection well permit process, perceptions of risk, reasons for involvement, involvement history, perception of governmental officials (trust). These first codes then began to cluster around patterns of mobilisation, respondents' social links and factors strengthening and weakening grassroots activity.

Interviews began in March 2009 and were conducted through to July 2010. As the case unfolded, all respondents were contacted at intervals of several months to assess changes in community action or reaction to the wastewater injection wells. Many other people were also involved (during 80 hours of ongoing participant observation) in meetings with residents, public meetings, public hearings and fundraising events. Notes from these conversations also inform this chapter.

The original focus of the research was community ability to organise and resist environmental hazards. As the residents described their involvement, however, they spoke about risk. It became clear that their perceptions of risks related to the wastewater injection wells motivated them to resist the wells. As the coding of the transcripts evolved, differences between how women and men explained their concerns about the wells surfaced. Although beyond the scope of this study, the data suggest that gender differences in risk perception may help explain the relatively large participation of women in collective action when toxic agents are involved.

This study is limited in several ways. First, it is based on a small case study of one area. Residents of other places may view risk differently from the people included in this study. Second, the research was initially centred on oppositional efforts; it did not involve women and men who were uninvolved in the fight against injection wells or ask about their views of risk. Non-participant residents may or may not see risk in other ways. Further study is needed to determine if there are fundamental differences in risk perception between local activists and residents who did not speak out against the wells. However, what does emerge from this study is important in several ways. First, these women's views on risk help explain their involvement in the grassroots opposition to the wastewater injection wells. To the extent that other women may have similar risk perceptions, these help us understand motivations to participate in grassroots activity when faced with environmental hazards. Second, these women's views on commercial wastewater injection wells suggest a vital area for additional research on how gender influences risk perception and risk management. Finally, men talked about risk in more formalised ways. Group members indicated that the ability of the men involved to communicate concerns in the technical language of the environmental agencies and administrative proceedings was vital in challenges to the wastewater injection well permits.

Men's views of risk

Most of the men involved in the case against the wastewater injection wells were linked to the cause as a result of their occupational roles as government employees, elected officials, legal counsel and technical experts. They did not live in neighbourhoods adjacent to the wells nor were they members of the grassroots group of residents. Their relationships to the injection wells and the struggle to resist them were more formalised and the language used to talk about them represented their legal, procedural and technical expertise in reports, legal documents, depositions and hearings.

However, several men were active in the core group of residents resisting the arrival of the wastewater injection wells, and therefore their relationships to the injection wells were comparable to those of the women involved in the opposition efforts. Even so, as they spoke of the wastewater injection wells and their concerns about possible risks, their language was much more technical than the women's. For example, when asked to explain the scheduling of a remand hearing by State Office of Administrative Hearing judges, one man explained it in this way: 'It was scheduled in July [2009] but the company got it delayed a little bit cause they were willing to re-perforate the well and do another draw-down test for the pressure value, to show how long it takes the pressure to drop off.' When asked to describe what this test would demonstrate, he continued:

> This is the porosity of the well, how far out it can spread toxic waste. It is very important. When the original owner of the injection well did the draw-down test for the pressure value they got like 89 millidarcies. When [the current owner of the injection well] put their application in, they did one at I think 500 millidarcies which is considered conservative.

Milladarcies, he explained, measure the porosity of the rock and how far the fluid would travel; a millidarcy is a unit of measurement of permeability. Lower millidarcies indicate a more porous substance, allowing increased flow of liquids. The man's concerns stemmed from discrepancies between the company's permit application information and calculations done by another member of the grassroots group to verify the company's claims about fluid migration:

> They just kind of did an average and we caught it and said 'Hey, you can't do this …'. The original test showed 89 millidarcies. [That] would make it spread out a whole lot farther than what they said it was going to be … even during the hearing. We asked our lawyers to ask the [company's] engineer and she said 'Well, the numbers just didn't add up' so we kind of went back and I plugged it in to our program and it came out. They were saying that it was going to spread out something like 3000 feet [915 m] and when she put the numbers in she said something like 10–15 000 feet [3000–4500 m] which was just astronomical. When asked whether she went back and checked her numbers she said 'No'. She said, 'I just dismissed it because it was so far from everything else that's been submitted.' You know that's incompetence right there that should have been a red flag to check on something like that because this is very important.

He explained the importance of the data on fluid migration. 'You need to have an idea of how far the fluids might travel, especially since there are fault lines in the area that could allow fluids to migrate upward into the aquifer that serves 54 Texas counties.' Fault lines did not represent the only risk, as another male respondent noted.

Another man, also a member of the resident opposition to the wastewater injection wells, told of his reasons for participating. He heard about the wastewater injection wells from a neighbour and fellow member of the property owners' association in his subdivision. Because of this respondent's technical background, the neighbour asked him to review the residents' concerns about the wastewater injection wells. He agreed to do so and concluded that the regulating agency, the Texas Commission on Environmental Quality (TCEQ), would not take the residents' objections seriously without a technical study. A comparison between the company's injection well permit application and the requirements of state law revealed 'various issues of noncompliance'. He detailed these in a technical report, which was submitted to the state to support opposition to the injection wells.

The man's concerns about the wastewater injection wells were linked to the history of the local oil and gas industry. The proposed wastewater injection well site was located in the old Conroe oil field that began producing oil and gas in the early 1930s. Over the years a number of wells were abandoned when their production declined or ceased. Some of these wells were located adjacent to the proposed injection well site. He asked:

> How in the world could TCEQ be ready to issue permits when [company's] application itself identified wells that they could find no records for? [The company] is proposing to inject waste into the Cockfield formation. They identified the Cockfield formation as their injection zone. Well, this was the same formation that was the producing formation in the Conroe Oil Field so oil and gas was being produced [from it]. So, I'm thinking here are all these abandoned wells that they can find no records for.

The presence of numerous ageing, abandoned wells and the lack of information about many of them contributed to his conviction that the wastewater injection wells were risky. He cited the testimony of another man involved, who:

> had worked in the Conroe Oil Field many years ago and had actually been involved in working on crews that were abandoning wells and some of the wells, he testified, they were told to just remove the 'Christmas tree', that's the valves and whatnot that are attached to the casing flange, just remove that and put a blind flange[6] on the casing and nothing was done to actually plug the wells.

To him, abandoned wells in proximity to the injection wells involved significant risk that the wastewater could contaminate the aquifers. He explained that, when wells are abandoned:

> casings are supposed to be plugged with what are typically referred to as 'mud plugs' and the state has specifications for the type of mud that's to be used – the density of the mud and so forth – so that's how the wells are supposed to be plugged so it seals off. When a well is sealed properly it is sealing off the well from the bottom all the way up to the surface ... When a well is abandoned, the producer is supposed to come in and actually plug the 'producing string' with that mud ... To prevent anything down in the formation, if there is some pressure build up, you don't want saltwater, oil or gas coming back up and contaminating anything.[7]

In his view, injecting wastewater at high pressure into the formations from which oil and gas were extracted many years ago could result in wastewater flowing back up and into the aquifers through old production wells that had not been properly sealed (plugged). The old metal casings could be corroded or cracked and could leak wastewater into underground sources of drinking water. The man argued that, for these reasons, abandoned wells, some within 150 feet (45 m) of the proposed injection wells, posed a real hazard. However, even wells that had been plugged properly could pose a threat and the old Conroe oil field was home to hundreds of old wells:

> I already knew that mud plugs don't always give you a 100% seal so my concern was, and I did all the math to prove this, that even if you had a mud plug with the density required by the state, even *if* you had a mud plug, the company is proposing to inject [wastewater] at between 1200 and 1500 PSI.

He explained that the pressure reported in the company's application was surface pressure that did not account for the additional pressure of a column of water from the surface to the injection interval some 5000 feet (1525 m) underground:

> The pressure at the injection interval was sufficient to overcome the pressure of a mud plug and through corroded casing you could have material that comes back up through abandoned wells, with or without mud plugs, and into the aquifers.

This man viewed the risk posed by wastewater injection wells through the lens of his technical expertise: 'I proved through mathematics (engineering) that the company could indeed contaminate our aquifers through their proposed system and that led to the contested case hearing.'

Framing the concerns of all residents involved, men and women, in the technical language used by the proponents of the wells, the state environmental agency and the legal representatives for all parties translated the views voiced by the women into more 'acceptable' and 'valid' terms. They spoke of the risk in terms that seemed to show the certainty of risk. The residents resisting the injection wells recognised that this man's contribution had been invaluable to the group's efforts to win a contested case hearing to publicly and legally challenge the wastewater injection well permits.

Women's views of risk

Although women are often absent from disaster literature or are portrayed as passive victims (Enarson *et al.* 2006), this research saw them at the centre of community efforts to limit the possibility of environmental disasters. The community study revealed that a perceived threat led women into social roles similar to those reported by Fothergill (2004). Like the women described by Brown and Ferguson (1995), women in this case study played key roles in opposing perceived environmental hazards and protesting toxic waste contamination.

The fact that more women than men become involved in grassroots efforts to resist the wastewater injection wells may indicate women's increased concern. Previous studies of gender and risk perception of environmental hazards indicated that women perceived hazards as more unacceptable and threatening than did men (Gutteling and Wiegman 1993). The women in this study were pessimistic about what they felt were unrealistic assertions by people (men) in favour of the injection wells, that the technology was fail-safe and that technical knowledge was trustworthy.[8] Interestingly, Flynn *et al.* (1994) found that white men perceived risks as much smaller than did people in every other category tested. Since gender differences in risk perceptions were significant only between white males and white females, variables such as power, status and trust in institutions seemed to contribute to the ways in which risk is perceived (Flynn *et al.* 1994). In this study, white men were present on both sides of the wastewater injection well issue, although all the proponents of the injection wells were white men and most of the residents opposing them were white women.

The majority of the core grassroots group members were women and several emerged as group leaders. These women talked about risk in different ways from the men involved. Three main differences emerged from the interviews. On the whole:

- women perceived uncertainty as riskier than men did;
- women talked about risk as linked to their roles as nurturers and caregivers while men spoke in terms of their technical/professional roles;
- women included a 'big picture' of risk, not just what might directly affect their lives or their families.

Each difference will be discussed in the following sections.

Risk and uncertainty

The word 'risk' refers to the possibility of loss or injury, or someone or something that creates or suggests a hazard (Meriam-Webster 2011). Noticeable in these definitions of risk are the words indicating uncertainty – 'possibility' and 'suggests.' As the women interviewed discussed their views on the commercial wastewater injection wells, they worried about the uncertainty of the consequences. 'Where will the wastewater really go?' 'Will the aquifer be safe?' 'Will the wastewater resurface?' The residents who carefully followed the legal proceedings heard experts who testified to the wells' safety and others who cautioned that they would be hazardous. What was the truth? If no one could know with certainty, these women argued, the wells were risky by definition. Uncertainty, in the case of the wells, was interpreted as risk.

For example, one of the first residents to become concerned about the risks posed by the wells explained that the company that owns them 'is trying to say that the pressure is going to push it to the bottom and it'll never come back up. But where there's a lot of pressure it's going to eventually come back to the top and if it does it can get in our drinking water.' Even if the 'plume' of the chemical–water mixture stays underground, she continued, 'you don't really, I don't know how they figure they *know* [where it is going to go].'

> They're saying that the plume wouldn't go out very much, and that's what I want to know. How do they *know* that? You know, when it goes down under the ground they don't know how far it's going spread. It could come up 10 miles [16 km] away!

Another woman living within half a mile (800 m) of the injection well site also had serious questions about the official version of the wells' impacts. She explained her doubts:

> Sometimes you fill up the bathtub with water and it's just yellow, like I don't even want to really get in that. I'm thinking, if that's happening with our well water already … what are they going do to our well water? [Will it] kill the animals? [Will there be] poison going through our skin? Because even taking a bath in what they're planning on putting in there [will be harmful].

She was not convinced by the company's presentation of how the wastewater injection well would function. She felt that the information provided by the company was insufficient:

> And another thing, they're saying it's mostly water. How do they know? At first they were saying that they were going to be diluting [the chemicals to be disposed of], or at least that's what everybody thought. Then we got to looking and they're not [going to be the ones] diluting it. They're depending on someone [else to] put it in that truck and they're going to claim to have diluted it. Will they dilute it? Who knows? They're going to bring it here and they're not even going to know what all they're putting together. You know, some chemicals can't mix. I don't care what they say about 'mostly water', you get a whole bunch of 'mostly water' and those chemicals get together, there is still going be a reaction somewhere, somehow. I want to see *them* drink it (laughs). That's what I want to do. Let's bring them a glass of it and let them drink it.

Another respondent also noted the problematic concept of 'mostly water' as she talked of past problems in trying to convince someone to clean up oil- and gas-related leaks in the area. She asked, 'What will happen when it is 'just water?'

> Where they're planning on putting [the wells] nobody's going to know until somebody dies or several people die before they make the connection that it's them. It would be a long time for somebody to prove that they caused it.

All the women interviewed worried about the uncertainty surrounding the wastewater injection wells. Some reported physical and emotional consequences, such as inability to sleep and severe anxiety. For example:

> There is such a huge part of me that is still sick to my stomach that … those wells could leak and people are going to get sick, whether it be in 10 years or 20 years.
>
> It's a lot. I think, it's just a lot to worry about.
>
> I've lost a lot of sleep, I've had a lot of anxiety over this and a lot of anger at just … the stupidity of why this has even gotten this far. Sometimes I felt like I was going have a heart attack just from stress.

Despite the uncertainty surrounding the possible impacts of the wastewater injection wells, the women were clear that they knew about risk. As they spoke about what they knew, women were quick to distance themselves from scientific or expert knowledge. In the fight against the wastewater injection wells, technical knowledge was the realm of male geologists, petroleum engineers, oil and gas industry workers and businessmen, traffic engineers, environmental experts and lawyers. The women clarified their knowledge with statements like, 'I'm not an engineer' or 'I'm no geologist'. They saw their knowledge as different and struggled to make their knowledge heard in the legal process. The women in this study relied on more subjective knowledge gained from personal experiences, gender roles and relationships to others (Belenky *et al.* 1986; Brown and Ferguson 1995).

Many of the women voiced healthy scepticism about official versions of the facts, knowing that facts are subject to interpretation. Drawing on experience, one woman pointed out that people and companies do not always follow the rules, especially when no one is watching:

> They monitor themselves! I mean, you don't know what's going to be put into that well. And you *know* they're not going to check every truck that goes in there to find out what's being put in that well.

All the women knew about the possibility that things do not always go as planned. For example, a woman explained that bad things can happen even in a very short time:

> It's still toxic chemicals being diluted so you would think that they would have a little more governing. The Texas Commission on Environmental Quality doesn't have the staff to go out there and check [the wells]. They '*try*' to check them once every couple of years. [In a] couple of years a lot of things can happen. [In] a couple of *days* things can happen.

Echoing other stories of sudden negative impacts of technology, another woman told how the death of a close family member when she was a teenager taught her about risk:

> [She] was killed in an industrial accident when I was 16 and instantly – my [relative] has been dead for 24 years – instantly [hearing about the wells] took me back. It's really because of what I went through, but the reality is all the information that I read, and I'm not an engineer, but *I truly believe in my heart* that that those wells [are going to contaminate the water].

Women's risk perceptions were informed by subjective knowledge gained through experiences, gender roles and relationships to others, and these views of risk motivated them to action (Brown and Ferguson 1995). Like that of the women in Brown and Ferguson's (1995) study, the gendered knowledge of these women struggled to find a place in the formal legal and procedural venues of the case against the wastewater injection wells. Their knowledge was considered neither pertinent nor admissible to the legal process, but it clearly guided women in their oppositional efforts. Their concerns could not find an effective voice until they were 'translated' by men into formalised technical language to 'prove' the risks that the women 'knew' were real.

Women's roles and risk

Uncertainty was understood by the women in this study through their social roles as caregivers and their relationships to others. The women most active in the wastewater injection well opposition represented a variety of roles – stay-at-home mother, housewife, secretary, beautician, estate sale organiser, golf tournament organiser, nurse, marketing professional. Women talked about the threat of the wastewater injection wells in terms of their roles as mothers, wives and community members. Motivations for involvement were linked to the protection of children, spouses, property values and community. Fear, frustration and anger related to their role as caregivers. They wondered how they could protect their loved ones from the negative impacts of the injection wells. In reaction to the hazard these permanent roles took on new meanings and new activities, and led to role strain as the women's time and energy were spread thin. Role strain refers to competing pressures associated with one role. For example, the role of mother entails caregiving for loved ones and (in the eyes of the female respondents) investing long hours researching, organising and protesting the wastewater injection wells – activities that take time away from caregiving:

> It's scary. It's frustrating. How come you're not seeing what I'm seeing? I looked at [the owner of the wastewater injection wells] at one of the hearings and I said, 'Look me in the eye and tell me 100% of those wells won't leak. Tell me.' He couldn't tell me that and I'm like, that's my point. *That's my point!* Why would I want to gamble my children's lives? Why would anyone?

A close relative of that speaker had died four months earlier from a mysterious illness related to chemical exposure. The speaker had been closely involved in her care and found it a traumatic experience. In her words, the illness was 'the most God-awful thing you can imagine'. This experience conditioned her reactions to the wastewater injection wells:

> Then all of a sudden here I am a couple months later and this stuff's coming up and I was terrified. I just kept thinking about when I walked out of that hospital I said I never want my children to have to go through that. It was horrible! And I vowed that my children would never have to go through this. [Now I'm hearing about the wastewater injection wells] and the reality is, my children may have to go through that. That's scary.

The recently deceased relative did not live in the same area, but her illness was the result of a benzene leak that contaminated the water in a Houston suburb. The type of hazard and the geographical location were both close enough to the risks of the wastewater injection wells to alarm the speaker. Her fears of chemical contamination, and suspicion of how authorities might handle it, were born from experience. She spoke with deep dismay, 'I don't know how it happens that way. I don't know how it happens'.

Another woman talked about her ill husband:

> Yes, and, you know, and that was one of the things that made me fight it so hard. He's already got cancer and if that stuff gets into your water, you don't know what it's going to do. It's not good, I can tell you that!

Several women spoke about concern for family members who live near the well site and who they feel will be at risk if the wastewater injection wells began operating:

> My son and his wife live right there within a half a mile [800 m] of the well. My granddaughter and her two children live there within a half a mile of the well. I have a sister and four nieces that live right there [with] their families. So there's a lot of family out there.

Another woman listed family members she was worried about:

> My oldest brother, my youngest brother, my mom, me. Then there's [my] aunt. [We all] live there right now. Then we have friends that are like family, we've all grown up together, we went to school with them [who also live adjacent to the well site].

Their concerns about potential impact of the wells led women to expand previous roles or take on new ones such as leaders, organisers, protesters, researchers, networkers, recruiters, fundraisers, public relations managers, public speakers, experts and activists. This led to role conflict

– the competing tasks, pressures and expectations experienced when holding more than one social role, e.g. being a mother and an environmental activist. These emergent public and political roles were key to the organisation of people and mobilisation of resources to resist the injection wells, but they came at a cost to the women – role conflict was mentioned frequently.

Women's willingness and ability to take on new roles in some cases meant personal, financial and social sacrifice for themselves and their families. Women spoke of the emotional toll of their activism with stories of conflict in relationships with neighbours, friends and family members. Others mentioned the strain on their finances. One woman lost her job because of the amount of time and energy she spent on opposition to the injection wells.

One woman's story illustrated the common issues of role strain and role conflict. She explained her involvement in terms of her roles as wife and mother and how those roles expanded as she became a leader of the group opposing the wells:

> Well, it's about our future also, you know, there's one thing when you're killing us off but when you start messing with our kids and our grandchildren people go, 'Wait a minute!' They take notice. The PTA [Parent–Teacher Association] was my resource because I'm very heavily active in the PTAs and in my kids' schools and all that volunteer work. That's where I went originally after I had gone to several of [the injection well] meetings and thought 'I need to pass this on'. So I started building through that.

As she took on the new and labour-intensive social roles of grassroots organiser, activist and group spokeswoman, she experienced role conflict. Her children even staged a makeshift protest at home to show their frustration that their mother's time and energy were so often focused on the opposition efforts. Other women also spoke of tension in relationships and finances, linked to role conflict.

Generalising risk

Women drew upon their personal experiences as caregivers generally, and with illness and death specifically, to urge neighbours, officials and policy-makers to be cautious about the wastewater injection wells. Their very personal experiences, however, gave them an ability to see a larger picture and a more generalised notion of risk. Some of the women lived close to the well site, so their actions could be understood in terms of self-preservation and protection of their interests. However, even these women spoke of the risk in broader terms, mentioning that the aquifer through which the wells would be drilled served 54 Texas counties. To them, it wasn't simply a local risk that could potentially harm only them. One woman summed up the group members' perspectives:

> This affects our water aquifers and if one of those wells gets a little crack or anything in it, it can ruin all of the water all the way to Houston, Beaumont. That's the reason we're fighting it so hard.

Another argued that protection of water is in everyone's interest:

> If there's nothing wrong with it, why do you have to pour it way down in the ground? That's why we're working on trying to protect the water. Our water is a precious commodity, you know, and if we run out of water, we're going to be in trouble.

Discussion

The men and women opposed to commercial wastewater injection wells in south-east Texas gave indications about how risk perceptions are gendered. It is clear from their stories that both men and women had gendered views of risk. The men understood and spoke of risk in very specific technical terms related to their occupations. The women were especially cautious in the face of uncertainty and viewed risks in terms of how it could endanger individual and community health. The women applied their personal experiences and subjective knowledge to a broader, more generalised concept of risk.

For the women involved in this case, uncertainty indicated risk. The link between uncertainty and risk in the minds and hearts of these women was not unique. Bord and O'Connor (1997) analysed the data from two national surveys and revealed that women were more concerned than men about specific risks. Other studies linked gender socialisation to women's increased feelings of vulnerability and heightened risk perception (Barke *et al.* 1997). Women's greater perception of risk was explained by suggesting that 'respondents who feel vulnerable answer in ways that urge caution on policy-makers' (Bord and O'Connor 1997, p. 830).

In this case study, risk was understood in relation to gender roles. The men relied on their technical expertise and the women on their roles of caregivers. In some cases, women were motivated to engage in new or increased activities linked to previous gender roles, causing role strain. Perceived risk also led women into new roles as community leaders. Women experienced empowerment and opportunity, but simultaneously faced role conflict if the new roles competed with other roles. Some of the women in this study described role conflict, and all experienced role strain. The men did not mention either.

In light of the disproportionate representation of male business owners, agency directors, politicians, lawyers, judges and technical experts involved in the case over the wastewater injection wells, grassroots groups attempting to challenge the permits had to overcome the obstacle of gendered risk perception. The men in the group were able to effectively translate the group's concerns into the technical language of the administrative hearings. In this way, the small group of residents was able to communicate with and convince local authorities that its concerns were valid.

As the case progressed, women began to adopt more technical language. This capacity to translate their concerns into more formalised language, combined with their ability to link personal experiences to a more generalised view of risk, were important assets that the women in this study brought to their new, more public roles. 'Men's higher risk thresholds are not necessarily more objective or more functional. The opposite may well be true' (Bord and O'Connor 1997, p. 834). This exploration of the risk perception of wastewater injection wells argues for greater incorporation of women's conceptualisations of risk in policy-making, particularly with regard to environmental hazards.

Endnotes

1 Water wells provide water to people living in areas without access to municipal water systems. Wastewater injection wells are repositories for diluted mixtures of chemicals from a variety of industrial processes. The wastewater mixtures may include antifreeze, acid, industrial waste, cleaning products like dioxins etc. The company applying for the injection well permits was not required to list specific substances it would inject underground.

2 In 2006, the EPA reported about 50 000 of these wells (EPA 2006).
3 Until 2011 this entity was named the Texas Railroad Commission.
4 The distinction between hazardous and non-hazardous waste was established under the *Land Disposal Program Flexibility Act 1996.*
5 Wastewaters may include ammonia, cyanide, magnesium, aluminum, nickel, phenols, sulfides, copper, iron, zinc, oil and grease, among other substances (Woodard and Curran 2006, p. 443).
6 The respondent explained that a blind flange is a flat steel plate bolted onto the mating flange of the casing to seal the casing.
7 Mud is the liquid circulated through the well bore during rotary drilling and workover operations. Mud protects against blowouts by holding back subsurface pressures, and deposits a mud cake on the wall of the borehole to prevent loss of fluids to the formation (Oil & Gas Glossary 2011).
8 Gwartney-Gibbs and Lach (1991) studied gender differences in perception of risk associated with nuclear technology.

References

Barke RP, Jenkins-Smith H and Slovic P (1997) Risk perceptions of men and women scientists. *Social Science Quarterly* **78**, 167–176.

Belenky MF, McVicker Clinchy B, Goldberger NR and Tarule JM (1986) *Women's Ways of Knowing: The Development of Self, Voice and Mind.* Basic Books, New York.

Berg B (2007) *Qualitative Research Methods for the Social Sciences.* 6th edn. Allyn and Bacon, New York.

Bord RJ and O'Connor RE (1997) The gender gap in environmental attitudes: the case of perceived vulnerability to risk. *Social Science Quarterly* **78**, 830–840.

Brown P and Ferguson FIT (1995) 'Making a big stink': women's work, women's relationships and toxic waste. *Activism, Gender and Society* **9**, 145–172.

Enarson E, Fothergill A and Peek L (2006) Gender and disaster: foundations and directions. In *Handbook of Disaster Research.* (Eds H Rodriguez, EL Quarantelli and RR Dynes) pp. 130–146. Springer, New York.

EPA (2001) Class I underground injection control program: study of the risks associated with Class I underground injection wells. US Environmental Protection Agency Office of Water, Washington.

EPA (2006) EPA injection well inventory. <http://water.epa.gov/type/groundwater/uic/upload/2007_12_12_uic_UIC-Inventory.pdf>.

Flynn J, Slovic P and Mertz CK (1994) Gender, race and perception of environmental health risks. *Risk Analysis* **14**, 1101–1108.

Fothergill A (2004) *Heads above Water: Gender, Class and Family in the Grand Forks Flood.* State University of New York, New York.

Gunter VJ and Kroll-Smith S (2007) *Volatile Places: A Sociology of Communities and Environmental Controversies.* Pine Forge Press, Thousand Oaks, California.

Gutteling JM and Wiegman O (1993) Gender-specific reactions to environmental hazards in the Netherlands. *Sex Roles* **28**, 433–447.

Gwartney-Gibbs PA and Lach DH (1991) Sex differences in attitudes toward nuclear war. *Journal of Peace Research* **28**, 161–174.

Meriam-Webster (2011) <www.meriam-webster.com/dictionary/>.

Oil & Gas Glossary (2011) Glossary. <http://oilgasglossary.com/mud.html>.
Strauss A and Corbin J (1990) *Basics of Qualitative Research: Grounded Theory Procedures and Techniques*. Sage, Newbury Park.
Woodard and Curran Inc. (2006) *Industrial Waste Treatment Handbook*. 2nd edn. Elsevier, Burlington.
Yin RK (2009) *Case Study Research: Design and Methods*. 4th edn. Sage, Washington.

8 DEVELOPMENT OF ENVIRONMENTAL RISK MANAGEMENT PLANS IN GREAT BARRIER REEF CATCHMENTS

Steve Purbrick and Nick Schofield

Introduction

Following its re-election in March 2009, the Bligh Labor government in Queensland committed to achieving ambitious water quality targets for the Great Barrier Reef (GBR). The proposed cuts in the amounts of pesticides, nutrients and sediments detected in GBR waters were to be achieved by imposing regulations on farmers in selected high-risk river catchments on the Australian mainland that flow into the GBR (GBR catchments). The backbone of these regulations would be Environmental Risk Management Plans (ERMPs) which, as the name suggests, use a risk-based approach in an attempt to reduce pollution in agricultural run-off and hence improve the health of GBR waters. This chapter details the development process of the risk framework that underpins the ERMPs.

The Queensland government targeted the Mackay-Whitsunday, Burdekin Dry Tropics and Wet Tropics regions for environmental risk planning as they carry the greatest pollutant loads to the GBR (Great Barrier Reef Protection Amendment Bill 2009 Explanatory Note). Specifically, this meant targeting the sugar and grazing industries: cane-growing properties larger than 80 ha in the Mackay-Whitsunday and Wet Tropics regions are required to prepare ERMPs under the regulations, as are grazing properties of more than 2000 ha in the Burdekin Dry Tropics (*Great Barrier Reef Protection Amendment Act 2009*). These different landscapes and types of agriculture have different impacts on the water quality of the GBR (Brodie *et al.* 2008). All three pollutants (pesticides, nutrients and sediments) may be discharged from agricultural properties, although the farming practices and thus methods of mobilisation for pollutants are industry-specific.

Socio-political context of ERMP regulations

The recent history of grazing and cane farming in Queensland does not lend itself to the simple imposition of regulations. Various legal and industry measures have been introduced in the last two decades, such as land management legislation and industry initiatives to encourage best management practices (Cavanagh 2000). A number of management planning frameworks preceded the ERMPs, including Land and Water Management Plans (which can be substituted for an ERMP), Soil Conservation Plans and Pest Management Plans. Land and Water Management Plans, for example, are required for the application of additional irrigation water and need to be approved by the Queensland government. ERMPs are therefore neither the first nor the only mandatory property-level management plans in use in Queensland.

Fluctuations in the prices for agricultural commodities, particularly sugar, mean that farmers already face a high degree of uncertainty in their industry (Sugar Industry Oversight Group 2004). For cane growers, the downstream impacts of their operations are generally only a secondary consideration to the viability of their properties (Kraack 2000).

Understanding of the impacts of agriculture on the GBR has improved over the last two decades and has been accompanied by a series of attempts at self-regulation and adoption of best management practices by industry (Rolfe *et al.* 2008). Despite these measures, the water quality of the GBR is still regarded as a threat to the Reef itself. This has resulted in the Queensland government legally requiring operators to develop management plans to reduce the risk of off-site water quality impacts (*Great Barrier Reef Protection Amendment Act 2009*). There is significant resistance from the sugarcane and grazing industries to the imposition of the new regulations. The challenge for the Queensland government is not just to improve water quality in the GBR, but to gain acceptance for its regulated risk management approach.

The Reef Plan water quality targets for the GBR are a 50% reduction in the amount of pesticides and nutrients flowing to the GBR by 2013 and a 20% reduction in the amount of sediments flowing to the GBR by 2020 (Dept of Premier and Cabinet 2009). The Queensland government approach to regulation is a key strategy under the Reef Plan to achieving these targets for GBR water quality. It may require significant changes in the way agriculture is conducted in those GBR catchments that discharge into the GBR. While adopting legally enforceable regulations will compel farmers to submit a risk management plan, there are broader questions about whether this approach is appropriate and whether the risk framework that underpins it will work as intended. Entrenching and building acceptance of the new approach will determine the success or failure of the regulations, although the risk framework still must be robust.

Scientific and economic context

Extensive research has gone into determining best practice methods for applying pesticides and fertilisers that have guided attempts at industry self-regulation (Rolfe *et al.* 2008). Such research is valuable, as operators can potentially make more profit by minimising the quantities of chemicals applied to their properties (Shannon 2002). As discussed later, one of the expectations underpinning this project is that farmers who were already following best practice principles for applying fertilisers and pesticides and for managing erosion and sediment run-off would not find it necessary to change their farming practices (Great Barrier Reef Protection Amendment Bill 2009 Explanatory Note).

There is a strong economic case for addressing the water quality of the GBR, with tourism estimated to be worth $5.8 billion each year to the Queensland economy (Access Economics 2007). Reef tourism is estimated to support 63 000 jobs (Access Economics 2007), meaning that there are considerable political, economic and social implications for the Queensland government if the number of tourists visiting the GBR declines. This growth in the tourism industry has meant that the health of the GBR is no longer just a debate about the financial concerns of graziers and cane farmers, but must also take into account people working in the hospitality and tourism sectors.

Threats to the GBR

The presence of pesticides and elevated levels of nutrients and sediments found in the waters of the GBR are largely attributed to the agricultural sector of the adjacent GBR catchments (Brodie *et al.* 2008). In cane farming, pesticides and fertilisers are applied to minimise the growth of weeds and to extract the largest sugar yields from each block (Faithful *et al.* 2007). In grazing, pesticides are typically used for managing woody weeds (Bainbridge *et al.* 2007). Sediment loads are generated by both industries due to soil erosion and the clearance of native vegetation (Rayment 2002).

A number of factors influence the discharge of pesticides, nutrients and sediments from farms. Variables such as the quantities of pesticides or nutrients applied, the timing of application (particularly with respect to rainfall), method of application, soil type and residual chemicals in the soil all influence what stays on-farm and what is discharged in run-off (Sinclair Knight Merz 2009). For example, higher levels of nutrients will be found in agricultural discharge if fertiliser has been applied above the requirements of sugar cane. Similarly, greater quantities of pesticides will be found in agricultural run-off if an application is followed by rainfall (Sinclair Knight Merz 2009).

The most important factor affecting pollutant discharge is the rainfall and run-off that transports pollutants off farms, into waterways and eventually to the GBR. The high level of rainfall in the GBR catchments is effective at moving pollutants from the soil and into waterways (Reghenzani and Armour 2000). The magnitude of rainfall during the northern wet season means that there are regular opportunities for pollutants to enter the GBR waters. Furthermore, some chemicals bond well to soil particles, which in turn carry chemicals out to the GBR (Haynes and Michalek-Wagner 2000).

Once pollutants reach the GBR, they affect the marine ecosystems in different ways. The pesticides most frequently found in GBR waters are herbicides, which affect aquatic plants. This in turn affects ecological communities such as seagrass beds, which provide habitat and food for sea animals (Haynes *et al.* 2000). The nutrients that have the most significant impact on the GBR are nitrogen and, to a lesser extent, phosphorus, from fertilisers. These nutrients are a major contributing factor to algal blooms, which can have a negative impact on corals (McCook 1999). For example, blue-green algae release toxins that are hazardous to a number of aquatic animal species (Devlin and Brodie 2005). Sediments tend to affect GBR ecosystems by blocking out sunlight, affecting the ability of aquatic plants and coral to produce energy via photosynthesis (Haynes *et al.* 2000). This can be due to the settling of larger sediments on coral reefs or seagrass beds, or to high levels of suspended sediments. This can have a significant impact on the health of the GBR, owing to the importance of coral in GBR ecosystems. Finally, sediments can provide an additional mechanism for the transport of chemicals that adsorb to sediment particles (Haynes *et al.* 2000).

In addition to the threats from agricultural run-off, the GBR is at risk from climate change. Two of the key climate change-related threats to the GBR, in particular coral, are sea temperature increases and ocean acidification. An increase in ocean temperatures as a consequence of global warming will increase the occurrence of bleaching events, where warm ocean waters kill coral, leaving only the bleached coral skeletons (Hughes *et al.* 2003). Increases in the amount of dissolved carbon dioxide also pose a threat to coral species, owing to the accompanying increase in the acidity of ocean waters (Hoegh-Guldberg *et al.* 2007). While it should be pointed out that large volumes of sediments were transported to the GBR prior to colonisation, the increase in sediment volumes and the presence of pesticides and nutrients are likely to hinder the ability of GBR ecosystems to cope with these climate change threats (Brodie *et al.* 2008).

First iteration of the risk framework

The Australian Standard

The starting point for building a risk framework was the Australian Standard for Risk Management (Standards Australia and Standards New Zealand 2004). The standard provides a robust method for identifying and assessing risk, and its key components were applied in the following way:

- identification of risks to the water quality of the GBR;
- assessment of the hazard posed by each risk (e.g. the amount of active ingredient in a herbicide);
- assessment of the consequences of each risk;
- assessment of the likelihood of each risk occurring;
- treatment of risks if level of risk is unacceptable.

If a numerical scoring system is used, the risk rating is determined by multiplying the scores for hazard, likelihood and consequence. A very low score for either likelihood or consequence will give a low risk ranking, with hazards given an implicit high score. The implication is that risks can be managed by minimising or completely eliminating the hazard, likelihood or consequence associated with each risk. Complete elimination of a hazard in this case could only be achieved by ceasing the application of pesticides and fertilisers. As a hypothetical example, the risk posed by polluted agricultural run-off could be effectively managed by managing water leaving each property. Given that each property is required to prepare an ERMP, risks can be managed at a property scale (as per the relevant legislation).

Risk management concept

In terms of managing water quality risks to the GBR, the risks are that pesticides, nutrients or sediments will be discharged to the GBR and damage its ecosystem. At a farm scale, every potential source of these three types of pollutants in the natural environment is a threat to the health of the GBR, as there is the potential for discharge to the GBR. The consequence of pesticides, nutrients or sediments finding their way to the GBR is assumed to be very high in all circumstances under the risk framework, as this allows for the protection of all species living within GBR waters.

With a high consequence ranking associated with the discharge of all three pollutants, this first model offers two ways of reducing risk. Risk can be reduced or managed by addressing the likelihood of each hazard occurring, or by reducing the hazard itself. Options for reducing the hazard are mainly dependent on input controls, such as changing the type of fertiliser used or the way it is applied, or reducing the amount applied. For reducing the likelihood, on-farm transport mechanisms need to be addressed via options such as undertaking soil conservation works, riparian planting or building tailwater retention dams. This reduces the discharge of pesticides, nutrients and sediments off farms and into waterways, which provide a means of transport to the GBR.

This approach was the basis for the first iteration of the ERMPs, where farmers would identify and rate risks before identifying appropriate management options. Under the conceptual model developed for considering risk, it is assumed that once pollutants are transported off-farm via a waterway they will arrive in the GBR at some point. The basis for this assumption is that all the GBR catchments have a high degree of hydrological connectivity to the ocean. The levels of rainfall in these catchments during the wet season further contribute to this connectivity; large run-off events mean that if pesticides, nutrients or sediments are present in the waterways, a high proportion will be discharged into GBR waters (Alibert *et al.* 2003). Any risk management actions taken by farmers therefore needed to reduce the availability of pollutants on their properties, or prevent polluted water from leaving their properties.

Specifics of the risk approach

Using the risk management standard as a guide, the principles of the initial risk management framework were based on the magnitude of the hazard present on the farm, the likelihood of transport from the farm to the GBR and the value of the asset. This can be represented as (Sinclair Knight Merz 2009):

$$\text{Risk} = \text{magnitude of hazard} \times \text{likelihood of transport} \times \text{value of asset}$$

The last aspect (value of the asset) is a means of prioritising risks if there are multiple assets. For example, two different assets may be exposed to the same level of risk. By taking the asset value into account, risk scores from different hazards and to different assets can be compared. It implies that if the asset has no value, then there is no risk. Values of an asset can be expressed in a variety of terms, such as ecological, economic and social values, and an asset may have multiple values.

At the level of planning required for the ERMP framework, however, the asset in question was the GBR as a whole, which is considered to have a very high value. As there are no realistic means of protecting the GBR asset from *in situ* exposure, this component of the risk framework was not given further consideration. The specific values of the GBR are identified above, but have not been investigated in great depth.

The 'likelihood of transport' component of the risk equation allows for variation in different subcatchments of the Burdekin Dry Tropics, Mackay-Whitsunday and Wet Tropics catchments. The time it takes pesticides, nutrients or sediments to reach the GBR will depend on the subcatchment, with some possessing longer pathways from farm to GBR. Location of a property within a subcatchment is also a factor in length of time between discharge of pollutants from farms and their arrival at the GBR. However, the high degree of hydrological connectivity between GBR catchments and the GBR itself means that although some pollutants may take longer to reach the GBR, they will reach it if discharged (Orpin *et al.* 1999). The extended transit time may allow for the breakdown of some chemicals, meaning that properties closer to the GBR may pose a higher risk (Sinclair Knight Merz 2009). However, a conservative approach assumes that all pollutants will be able to reach the GBR, regardless of transit time. The likelihood component of transport from the discharge point at the farm boundary to the GBR is therefore not evaluated.

This implies that managing the risk from pesticides, fertilisers and sediments needs to occur principally at the property level, which has two components: source (hazard) control and control of on-farm transport to the farm boundary discharge point. The hazard can be managed by controlling the availability of the pollutant (as an input or as the release/mobilisation of sediment) at its source. There are a number of factors that affect the availability of pollutants (hazard), including type and amount of chemical applied, timing and method of chemical application, presence and stability of erosion sites, and frequency and intensity of storms (Cavanagh *et al.* 1999; Reghenzani and Armour 2000). There are also a number of farming practices and additional options for controlling on-farm transport, including irrigation application and design (e.g. centre pivots, design of furrows, laser levelling), on-farm drainage design, recycling ponds, wetland filters, riparian grass and tree buffers.

First draft of the ERMP

For each property, farmers would be required to identify each potential source of a pollutant, whether it was an erosion hotspot, application of pesticides or fertilisers and storage of pesticides or fertilisers. Each hazard would be scored on a scale of 1–5, based on its type and quantity. This combination would be different for each type of pollutant and would vary according to the practices of different farmers. At this stage of the assessment it was important that farmers were provided with clear instructions on how to score risks. For example, the relationship between quantities of fertiliser or pesticide applied and risk rating was subject to a number of variables, including some idea of what constituted 'ideal' or 'best practice' methods. It became clear that the successful implementation of the ERMP framework required clear and perhaps extensive assistance for the affected industries. Once drafted, ERMPs would be assessed by the Queensland government to ensure rigour and compliance with legislation.

Farmers would then be required to make an assessment of the likelihood of agricultural run-off leaving their properties. Like the assessment of the previous step, this would be on a scale of 1–5 and it too required some guidance on relating an abstract numerical rank to the likelihood of discharge of agricultural run-off. Factors that might affect this aspect of risk assessment include irrigation practices, frequency and intensity of storms/rainfall, location and topography of property, soil properties, presence of run-off retention dams and the flood zones of rivers (Simpson *et al.* 2000).

Hazard and likelihood of discharge scores would be multiplied to give the risk score for each property. The total score would indicate the level of risk and determine whether a management response is required. On the 1–5 scale, 3 is considered a medium score (where 1 = very low, 2 = low, 4 = high and 5 = very high). Thus an overall score of 9 is equivalent to two medium scores, and anything less means that at least one of the hazard or likelihood scores was low or very low.

For each risk identified, farmers would be required to nominate a management response that would appropriately reduce the risk. In the scheme described, farmers could focus on hazard or transport control, or both, to reduce risks. To account for existing work on reducing water quality risks, an assessment was made of the alignment between industry best practices and agricultural extension activities with the demands of the ERMP framework. Initiatives from groups such as Meat and Livestock Australia, natural resource management groups and BSES Ltd (formerly Bureau of Sugar Experimental Stations) were considered, to assess whether they could partially or fully replace aspects of the ERMP. Those that used a robust risk-based approach were considered as potential substitutes for the Queensland government's ERMP and could be assessed as an alternative document equivalent to an ERMP. Those that provided guidance on management actions were considered as a source of risk management responses. The intention was to provide farmers with a variety of options and advice. Once completed, ERMPs would be accredited by the Queensland government, meaning that risk assessments and risk responses were reviewed to identify the extent of risk to GBR water quality and compliance with the intent of the *Great Barrier Reef Protection Amendment Act 2009.*

Second iteration of the risk approach

Simplifications for practical implementation

The second iteration of the risk framework attempted to maintain the risk approach developed in the first ERMP model, while adopting a more user-friendly approach. This was a response to the potential difficulties with requiring farmers to adopt a quantitative risk assessment method. Other problems with this approach included its focus on compliance with a process rather than on the improvement of management practices. If the ultimate goal of the ERMP process is to improve water quality by improving farming practices, then the ERMPs need to educate and assist farmers as much as possible. However, in order to adhere to the risk principles of the initial concept, the second iteration still sought to address risks by reducing the hazard or reducing the options for the discharge of agricultural run-off.

The second iteration of the ERMPs attempted to simplify some of the complexity of the initial concept. Rather than requiring farmers to quantitatively rank the risks their agricultural run-off posed to the GBR, the ERMP was simplified to a series of yes/no questions that identified the potential sources of water quality pollutants. It broke the risk assessment stage down into three sets of questions covering sediment, nutrient and pesticide pollution. These sections addressed the inputs of pesticides or fertilisers, the availability of each pollutant and the likelihood of it being carried off-farm by agricultural run-off. 'Inputs' refer to the types and quantities of a chemical applied, hence the exclusion of sediments from this category. 'Availability' refers to the ability of water to pick up pollutants and thus covers considerations such as application method or, in the case of sediments, the presence of areas of eroding soil. 'Transport' refers to mechanisms that allow water to carry any pollutant off-site and presumably into waterways that reach the GBR. For each question, yes or no answers corresponded with a high or low risk rating. If a 'high' were scored, farmers would be required to address the activities identified as high-risk.

This approach, while an attempt to simplify the original ERMP concept, required a more intensive approach to risk management. Under the previous model, only one aspect of each risk demanded a management response (e.g. restricting the off-farm discharge of agricultural run-off). Under this second approach, a risk management response was required for each component of each risk (e.g. restricting off-farm discharge would not be sufficient; input and availability controls would also be required). This could create inefficiencies, such as requiring transport controls on a property that had already minimised the availability of pollutants. In this case, such a measure would clearly be unwarranted from a risk management perspective, as risk had already been minimised.

This model therefore made an attempt to identify practices that were inherently high-risk and required farmers to move towards those identified as best practice. Although this removed some of the guesswork involved in filling out the ERMP, it did assume that all risky practices had been identified within the ERMP questions. It also presumed to have identified the practices essential for lowering the impact of agriculture on the GBR. For example, the series of questions on transport was largely the same for the pesticide, nutrient and sediment assessments, and regarded the lack of vegetative buffer strips on cane properties as high-risk. Such a control, however, may do very little for minimising off-site water quality risks after storms (Reghenzani and Armour 2000). The importance of a comprehensive and accurate set of questions was therefore critical to this approach.

Other implementation aspects

At this stage of the project, a range of recommended industry tools was provided to assist farmers to develop risk management controls. The tools largely consisted of a range of best practice-type programs, and were identified separately for the sugarcane and grazing industries. The programs and tools were reviewed for alignment with the needs of the ERMP approach. This led to the identification of programs that could guide farmers on possible changes to their farm operations.

In addition to the risk assessment, farmers were required to provide a property map as part of their ERMP. This was particularly important for assessing the flow of water around and across properties – a key factor in pesticides, nutrients and sediments ending up in the GBR. Properties that controlled agricultural run-off and flooding risk were considered to be reducing a significant aspect of their risk to the GBR (Dept of Environment and Resource

Management 2010a) and the maps allowed the Queensland government to assess if this were the case. Other claims made in the ERMP could be tested through records such as pesticide or fertiliser application logs.

A key substitute for an ERMP is a Land and Water Management Plan (LWMP). These plans, an initiative of the Queensland government, committed farmers to certain practices that reduced the impact of irrigation tailwater on land and water bodies (Dept of Environment and Resource Management 2010b). Given the emphasis these plans placed on soil and water health, they were considered an appropriate tool for committing farmers to practices that minimised off-site risks to water quality. An LWMP could be submitted in place of and be approved as an ERMP provided it met the criteria of an ERMP and covered the whole extent of the property (*Great Barrier Reef Protection Amendment Act 2009*).

This was the point at which SKM's formal involvement in the project ended; final development was carried out by the Queensland government in consultation with user groups. The final versions of the sugarcane and grazing ERMPs can be found at http://www.reefwisefarming.qld.gov.au/information/ermps.html and the first ERMPs were due by the end of September 2010. The length of time between the commencement of the new planning regime and improvements to GBR water quality will vary according to location. It should be noted that, under the *Great Barrier Reef Protection Amendment Act 2009*, farmers in the three affected GBR catchments were required to restrict fertiliser application to optimum amounts after 1 January 2010. This means that the amount of nutrients flowing towards or into the GBR should have begun to decrease. Water quality and ecological monitoring will be required for a number of years to determine the effectiveness of the new regulations for GBR health.

Discussion

From a risk management perspective, the environmental risk management planning framework in its final form addressed the key elements of risk management – the availability of pesticides, nutrients or sediments, and the ability of these pollutants to be transported to the GBR (the final ERMPs contained a section on 'Water run-off management'). High-risk management practices were identified, with controls nominated in the action plan section. The final ERMP was an example of an applied risk management framework which attempted to take a user-friendly approach to environmental risk management. The effect of using a regulatory approach for improving the water quality of the GBR will become more apparent as time progresses.

The ERMP framework adopted by the Queensland government can therefore be considered as an applied set of risk management principles. The key aspects of risk, consequence and likelihood were identified and managed for agricultural run-off in the GBR catchments. In particular, the framework addressed the components of risk identified in the first ERMP iteration. The sources of pesticides, nutrients and sediments were identified, as were the mechanisms for their discharge in agricultural run-off. The structure of the ERMP now lends itself to changes in agricultural practices, with high-risk practices identified and potential solutions nominated to assist farmers with completing their ERMPs.

Conclusion

Only time will provide sufficient data to determine if the Queensland government's decision to regulate the agricultural sector using an environmental risk management approach has been effective. From a scientific perspective, risk management allows the optimum targeting of improvements in the quality and quantity of agricultural run-off, with regulations to ensure enforcement of this approach. Assuming that there have been no significant oversights in the design of the risk framework, the success of this approach hinges on its implementation. This requires participation by farmers in the affected GBR catchments, who need to accept the value of the GBR and the impact of agricultural run-off, and accept that the solution lies in altering farming practices. Cottrell (2000) identified a number of social factors that are potential barriers to the adoption of more sustainable practices in the sugar industry, and argued for governments to provide financial support where new management practices are required. If inadequate compliance and enforcement, support or information is provided to farmers, the Queensland government will have difficulty effecting changes in the behaviour of sugarcane growers and graziers. There are a number of other contributing factors, such as the attitudes of industry bodies, availability of Queensland government staff and access to a process that allows farmers to air grievances. The approach needs to be long-term and it requires significant changes from the status quo. The ERMP regime will need to be in place for a number of years before its effectiveness can be assessed, and any changes it brings need to be permanent.

Acknowledgements

The authors would like to acknowledge the assistance and contributions of the project team, including BSES Ltd and Dr Michael Quirk. The work of SKM project manager Anne Chan, and Jean Erbacher at the Department of Environment and Resource Management, is particularly appreciated.

Endnote

Disclaimer: this chapter represents the personal views of the authors and does not represent the opinions of the Queensland government. Furthermore, the current policy environment does not necessarily reflect the environment in which the environmental risk management plans were developed.

References

Access Economics (2007) Measuring the economic and financial value of the Great Barrier Reef Marine Park, 2005–2006. GBRMPA Published Report No. 84. Great Barrier Reef Marine Park Authority, Townsville.

Alibert C, Kinsley L, Fallon S, McCulloch MT, Berkelmans R and McAllister F (2003) Source of trace element variability in Great Barrier Reef corals affected by the Burdekin flood plumes. *Geochimica et Cosmochimica Acta* **67**, 231–246.

Bainbridge Z, Lewis S and Brodie J (2007) Sediment and nutrient exports for the Burdekin River catchment, north queensland: a comparison of monitoring and modelling data. In *Proceedings of the MODSIM 2007 International Congress on Modelling and Simulation.* 10–13 December 2007, Christchurch. (Eds L Oxley and D Kulasiri). Modelling and Simulation Society of Australia and New Zealand.

Brodie J, Binney J, Fabricius K, Gordon I, Hoegh-Guldberg O, Hunter H, O'Reagain P, Pearson R, Quirk M, Thorburn P, Waterhouse J, Webster I and Wilkinson S (2008) Scientific consensus statement on water quality in the Great Barrier Reef. Reef Water Quality Protection Plan Secretariat, Qld Dept of Premier and Cabinet, Brisbane.

Cavanagh JE (2000) Comparison of the environmental regulation of land management in the sugar industry under the *Sugar Industry Act 1991* (Qld), the Sugar Industry Bill 1999 and

the *Integrated Planning Act 1997* (Qld). *Environmental and Planning Law Journal* **17**, 118–125.
Cottrell A (2000) Sociological considerations in fostering change and achieving adoption of best practice: an overview. In *Environmental Short Course for Sustainable Sugar Production: Course Manual*. (Eds RC Bruce, M Johnston and GE Rayment) pp. 29–33. CRC for Sustainable Sugar Production, Townsville.
Dept of Environment and Resource Management (2010a) Guide: sugarcane growing Environmental Risk Management Plan (ERMP). Qld Dept of Environment and Resource Management, Brisbane.
Dept of Environment and Resource Management (2010b) Land and Water Management Plans. Qld Dept of Environment and Resource Management, Brisbane.
Dept of Premier and Cabinet (2009) Reef Water Quality Protection Plan 2009 for the Great Barrier Reef World Heritage Area and adjacent catchments. Reef Water Quality Protection Plan Secretariat, Qld Dept of Premier and Cabinet, Brisbane.
Devlin MJ and Brodie J (2005) Terrestrial discharge into the Great Barrier Reef lagoon: nutrient behavior in coastal waters. *Marine Pollution Bulletin* **51**, 9–22.
Faithful J, Liessmann L, Brodie J, Ledee E, Sydes D and Maughan M (2007) Water quality characteristics of waters draining different land uses in the Tully/Murray rivers region. ACTFR Report No. 06/25, February 2007. James Cook University, Townsville.
Great Barrier Reef Protection Amendment Act 2009 (Qld).
Great Barrier Reef Protection Amendment Bill 2009 Explanatory Note.
Haynes D and Michalek-Wagner K (2000) Water quality in the Great Barrier Reef World Heritage Area: past perspectives, current issues and new research directions. *Marine Pollution Bulletin* **41**, 428–434.
Haynes D, Müller J and Carters S (2000) Pesticide and herbicide residues in sediments and seagrasses from the Great Barrier Reef World Heritage Area and Queensland coast. *Marine Pollution Bulletin* **41**, 279–287.
Hoegh-Guldberg O, Mumby PJ, Hooten AJ, Steneck RS, Greenfield O, Gomez E, Harvell CD, Sale PF, Edwards AJ, Caldeira K, Knowlton N, Eakin CM, Iglesias-Prieto R, Muthiga N, Bradbury RH, Dubi A and Hatziolos ME (2007) Coral reefs under rapid climate change and ocean acidification. *Science* **318**, 5857, 1737–1742.
Hughes TP, Baird AH, Bellwood DR, Card M, Connolly SR, Folke C, Grosberg R, Hoegh-Guldberg O, Jackson JBC, Kleypas J, Lough JM, Marshall P, Nyström M, Palumbi SR, Pandolfi JM, Rosen B and Roughgarden J (2003) Climate change, human impacts and the resilience of coral reefs. *Science* **301**, 5635, 929–933.
Kraack A (2000) Why do growers do what they do? Farming behaviour and attitudes in relation to crop, soil and water management. CRC Sugar Occasional Publication. CRC for Sustainable Sugar Production, Townsville.
McCook LJ (1999) Macroalgae, nutrients and phase shifts on coral reefs: scientific issues and management consequences for the Great Barrier Reef. *Coral Reefs* **18,** 357–367.
Orpin AR, Ridd PV and Stewart LK (1999) Assessment of the relative importance of major sediment transport mechanisms in the central Great Barrier Reef lagoon. *Australian Journal of Earth Sciences* **46,** 883–896.
Rayment GE (2002) Water quality pressures and status in sugar catchments. Exposure Draft. CRC for Sustainable Sugar Production, Townsville.
Reghenzani JR and Armour JD (2000) Management to minimise nutrient export from canelands. In *Environmental Short Course for Sustainable Sugar Production: Course*

Manual. (Eds RC Bruce, M Johnston and GE Rayment) pp. 117–122. CRC for Sustainable Sugar Production, Townsville.

Rolfe J, Wake J, Higham W, Windle J and Trendell P (2008) Best management practice adoption in the sugar industry for improving water quality: the Mackay-Whitsunday region. *Consortium for Integrated Resource Management Occasional Papers.* Qld Dept of Natural Resources and Water, Coorparoo.

Shannon G (2002) Optimisation of nutrient management of the Queensland sugar industry. Sugar Research and Development Corporation Final Report SD02006. SRDC Project BSS176. BSES Publication, June 2002.

Simpson BW, Hargreaves PA and Ruddle LJ (2000) Responsible pesticide use. In *Environmental Short Course for Sustainable Sugar Production: Course Manual.* (Eds RC Bruce, M Johnston and GE Rayment) pp. 123–127. CRC for Sustainable Sugar Production, Townsville.

Sinclair Knight Merz (2009) Reef Protection Package: draft ERMP concept design. Report to Qld Dept of Environment and Resource Management, Brisbane.

Standards Australia and Standards New Zealand (2004) Australian Standard/New Zealand Standard 4360:2004 Risk Management. Standards Australia, Sydney.

Sugar Industry Oversight Group (2004) The strategic vision for a commercially vibrant, sustainable and self-reliant industry in Australia. Report to the Dept of Agriculture, Fisheries and Forestry, Canberra.

9 SHIFTING SANDS: UNCERTAINTY AND A LOCAL COMMUNITY RESPONSE TO SEA LEVEL RISE POLICY IN AUSTRALIA

Anne Leitch and Cathy Robinson

Introduction

Sea level rise presents uncertain environmental and social conditions that require a rethink of institutional arrangements for the Australian coastline (Abel *et al.* 2011; Ryan *et al.* 2011). Coastal policies need to have a degree of public understanding of the required trade-offs to be successfully implemented (Tompkins *et al.* 2008). Therefore governments will need to balance the expectations, aspirations and values of coastal communities. Central to this balancing act is the need to account for the social dimensions of uncertainty which play an important role in the ability of local authorities to transition to new regulatory and legislative arrangements. In this chapter we use environmental planning theory to better understand concepts of uncertainty in one community on the east coast of Australia. We analyse the community discourse following the implementation of a local government policy for adapting to sea level rise. Through media analysis, we trace the varying perceptions of uncertainty articulated by community members and reflect on how these discourses help us understand and engage with local contexts and conflicts caused by complex environmental changes and policy responses.

Policies for managing erosion in developed coastal areas are being discussed by local governments. Planned retreat is becoming more common in the US and UK, and Byron Shire in New South Wales is the first local government area in Australia to implement a local policy of planned retreat. Policies of planned retreat acknowledge dynamic coastal processes as natural phenomena that are allowed to take their course rather than be managed through built structures or similar hard engineering options. In this study we explore community discourse through the local media during the implementation of a sea level rise policy in Byron Shire on the east coast of Australia, following a damaging coastal erosion event in May 2009. While there are numerous examples of media analyses of climate change at national scales (Boykoff 2008a, 2008b; Doulton and Brown 2009) it is rarely considered at the local scale or as local impacts. A local community discourse reveals the local understanding of the phenomenon, differing perspectives of evidence and knowledge and associated issues, different expectations of authority and action, and normative judgements of values, concerns, risk and uncertainty. We argue that ultimately more realistic, and maybe more useful, sea level rise policies will emerge if we have a better idea of the locally embedded realities and constraints that can influence community responses to sea level rise.

Local governments as managers of coastal areas

In Australia's three-tiered system of government, local governments are primarily responsible for managing sea level rise based on their extensive role in service delivery, land use and planning, and management of public assets (Australian Government 2010). In adapting to sea level rise, local governments in Australia face a number of complex and overlapping challenges. First, many coastal towns have soft, sandy coastlines which are particularly susceptible to erosion. Of these cities and towns, many already experience coastal erosion in developed areas (Harvey *et al.* 1999; Harvey and Clark 2007). This erosion is likely to be magnified and hastened by climate change with a projected sea level rise, relative to 1990 mean sea levels, of 40 cm by 2050 and 90 cm by 2100 (IPCC 2007). A rising sea is likely to be accompanied by increasing extreme weather (Walsh *et al.* 2004; IPCC 2007) resulting in more erosion events (Hunter 2009). For local governments in coastal areas there is increasing pressure to consider sea level rise in planning, as failing to prepare adequately may expose councils to criticism or 'voter dismay' in their constituents (Morrison *et al.* 2009), as well as legal liability (England 2006; McDonald 2007). Yet coastal councils are struggling to incorporate sea level rise into their planning (Morrison *et al.* 2009) with little statutory and regulatory guidance from state governments (Lipman and Stokes 2003; McDonald and England 2007).

A second challenge for local governments is that areas within their jurisdiction have been subject to increased development since the 1960s when a steady flow of population began moving to non-metropolitan centres (Burnley and Murphy 2004). Population growth in coastal areas places more people at risk from rising sea levels. Population influx can also place increasing pressure on local identity and character (Berwick 2007) and local environments (Harvey and Woodroffe 2008) as well as the local authority's ability to manage (Harvey and Caton 2003). Beach shacks, caravan parks and open spaces on coastal foreshores are frequently replaced with expensive residences built for home owners with a lot of resources and often a high capacity for political influence (McDonald 2007; Lyne 2010; Anon. 2010). The rapid population growth also results in increasing community diversity, which can mean divisions between existing and newly arrived residents (Gurran *et al.* 2005). Local communities are often highly heterogeneous and divided over a range of issues (Lane and McDonald 2005; Measham *et al.* 2011).

These two challenges are made more complex given that local governments face an antimony of power (Wild River 2006). Local government is a hybrid organisation: its form and function is shaped by state government yet every three to four years it is elected by the community it governs. This hybrid nature means that local authorities have boundaries, roles and responsibilities determined by the state government within its own political cycle and agenda (Binning *et al.* 1999; Dollery 2005). Within these state-set parameters, issues compete for attention in a contested and resource-limited planning agenda steered by local political interests (Measham *et al.*, in press; Wild River 2002). At the same time, councillors attempt to gain support, or at least consent, from their communities for a spectrum of issues as they seek re-election. The hybrid nature of local government presents challenges for many environmental issues (Wild River 2002) and it becomes especially difficult with a contested and complex issue such as sea level rise.

Planning for sea level rise and uncertainty in a contested environment

Increasing awareness of sea level rise and increasing community conflict over coastal erosion events highlight the inadequacy of current coastal planning institutions. Planning codes and guidelines often rely on predictability and are based on a presumption that experts can design workable solutions to meet recognised goals (Innes and Booher 2003). In response to these challenges, there are escalating calls for new institutions for coastal communities that can manage assumptions about uncertainty, incomplete control and multiple perspectives (Bray *et al.* 1997; Norman 2009; Morrison *et al.* 2009; Abel *et al.* 2011). Leitch *et al.* (2010) broadly outlined four types of uncertainty that are relevant in local planning response to climate change:

- scientific uncertainty due to limitations of current and relevant biophysical knowledge;
- epistemological uncertainty of whose knowledge and values count;
- ethical uncertainty of who is responsible and accountable;
- uncertainty of scale regarding action, roles and responsibilities.

The four types of uncertainty are not discrete but have substantial overlap. Here we reinterpret these four types of uncertainty and how they are understood at the community scale as it pertains to sea level rise.

Scientific uncertainty: the sky's the limit

Despite the relative confidence of global predictions, scientific knowledge of sea level rise remains uncertain, incomplete and contested (Aarup *et al.* 2010). For complex and contested environmental problems such as sea level rise, scientific knowledge – and frequently the 'best available science' – is a cornerstone of the debate (Wynne 2002). The science claims are used by some interests to justify their position or action and countered by other interests who dispute their credibility and/or certainty (Oreskes 2004; Sarewitz 2004). The lack of sufficient scientific information to make decisions at the local scale is often invoked as a reason for inaction (Moser and Dilling 2004; Moser 2005).

For climate change and its consequences, the limitations and local-level uncertainties associated with climate science are the dominant discourse in political and media debate (Weingart *et al.* 2000; Boykoff 2008a). At the global scale, the uncertainty debate for sea level rise involves thermal expansion of the oceans, regional landmass change and ice sheet flows and melts (Lowe and Gregory 2010), difficulty of direct measurement and different models and methods used to project impacts, and the scale of information available (Aarup *et al.* 2010). At the local scale, the uncertainty debate tends to emphasise different, more locally derived contests such as causes of erosion (Milligan and O'Riordan 2007; Cooper *et al.* 2009). There is a growing realisation that biophysical science provides only a subset of the knowledge needed to manage complex problems, and that other types of knowledge are necessary.

Epistemological uncertainty: what counts as knowledge and whose knowledge counts?

Adaptation to sea level rise requires local knowledge: there is a need to know what local people already understand and value about their system through contextually valid knowledge based on experience and reflection (Innes 1998; Wynne 1989; Healey 1999; Fischer 2000; Tompkins and Adger 2005; van Kerkhoff and Lebel 2006). Local knowledge can provide relevant information about the context of climate change impacts and responses as well as offering new information and ideas about climate change strategy design (Rydin 2003). Incorporating local knowledge can help to include a diversity of values into the debate (Healey 1999), may help to build consensus between different stakeholders (Innes 1996) and may facilitate innovation and learning within local communities (Failing *et al.* 2007). Local knowledge can influence individual decision-making, as historical knowledge or past experience of a location may affect an individual's decision to purchase coastal property (Turbott and Stewart 2006; Ryan *et al.* 2011; Alexander *et al.*, in press). Omitting or excluding local knowledge – especially if it conflicts with expert knowledge – may affect the credibility and legitimacy of policy process and decision-making (Wynne 1992; O'Riordan and Ward 1997).

The need to incorporate different types of knowledge raises wider issues of uncertainty about who should be involved in planning and about how power relations affect planning and the inclusion of values (Friedmann 1973; Forester 2007). The selection, integration and use of knowledge is not without the 'peculiarities of perspective and politics' (Lahsen 2007) and therefore knowledge 'brokers' such as local governments may choose what, and whose, knowledge to include or to ignore and what different knowledge types can contribute (Healey 2006; Innes 1998). Framing of particular viewpoints is influenced by the local context and power dynamics that can determine if and what knowledge is deemed credible or legitimate (Flyvberg 2001).

Ethical uncertainty: who bears the consequences?

Ethical uncertainty pertains to the equity of who and what affects, and is affected by, climate change impacts and response. While sea level rise threatens all coastal regions, there are some ecosystems, communities and subgroups at locations and within communities which will be more affected due to a spectrum of biophysical and social vulnerabilities (Tol *et al.* 2008; Dept of Climate Change 2009). In developed countries, it is often the well-resourced property owners who are affected by sea level rise (Cutter and Emrich 2006; Hayward 2008), resulting in conflicts over individual property rights. Ethical uncertainty extends beyond that concerned with individual property rights as different adaptation responses will have different biophysical impacts on the coast and some, such as hard engineering structures, can lead to loss of the beach, a highly valued public asset (Cooper and McKenna 2008). Intertwined with issues of power and local context, personal resources influence the socio-political dimensions of decision-making through the capacity of stakeholders to get involved, garner political support, form coalitions (Moser 2000) and access information or alternative experts (Fischer 2000). Depending on the adaptation response, the community may be more or less affected by having to bear the costs of response either directly or indirectly through expenditure of public funds.

Uncertainty of scales of responsibility: who should lead action?

Bulkeley and Betsill (2005, p. 43) described climate change as meshing 'the global and local in the presence of the nation-state', so generating complex multi-level governance issues. Planning for sea level rise takes place primarily within local government, the main managers of the coastal zone. State governments control the legislative basis for land use and coastal planning and have oversight of local government, which is responsible for developing its local land use and coastal plans within that legislative framework (Haward 1996). Typically, local councils conduct land use planning to manage development patterns in the coastal zone and attempt to accommodate the views, values and aspirations of disparate stakeholders (Allmendinger *et al.* 2002).The complexity of these governance issues is compounded by uncertainty about scale of action and consequences of response: policies designed to address issues at a regional scale can have unforeseen results at a local scale (Brooks *et al.* 2005). Conversely, a lack of policy guidance at the national scale can leave the local scale operating in a policy vacuum (O'Hagan and Ballinger 2010). This is likely to produce short-term thinking, lack of coordination and local community conflict. It also exposes local government to the threat of litigation.

These dimensions of uncertainty are evident from the analysis of community understandings and responses to sea level rise in a coastal community on Australia's north-east coast.

Implementing a policy of planned retreat in Byron Bay, NSW

Settled areas along the Australian coastline are complex policy arenas, governed by agencies with differing but often overlapping jurisdictions, objectives, responsibilities and powers

(Allmendinger *et al.* 2002; Nicholls 1999). The NSW government began to formally consider sea level rise in 1988, but it was two more decades before the state developed planning guidelines for sea level rise. In 2009, the state government released a draft sea level rise policy (largely in response to coastal erosion from recent storm events in several NSW areas) which included planning benchmarks of an increase of 0.4 m by 2050 and an increase of 0.9 m by 2100 (DECC&W 2009; Watson 2011). Local councils in NSW are in various stages of developing their own response to coastal erosion events through land use or coastal planning, or a combination of both. The potential of planned retreat as a management strategy was mooted as a policy of relocation for Byron Shire (PWD 1978) and included in the NSW government's *Coastline Planning Manual* in 1988.

Byron Shire is a coastal local government area in northern NSW, Australia. Dominated by tourism and agriculture, the local government area of Byron Shire is considered a 'sea change' coastal lifestyle destination (Gurran *et al.* 2005) with census data showing almost a tripling of the shire's population from approximately 11 000 to 30 000 residents between 1976 and 2006 (ABS 2006). House prices also increased dramatically in that time, with beachside houses recently selling for \$2–\$16 million (Johnstone 2010). Records show that since 1888 Byron Shire's beaches have experienced a number of storm events resulting in significant damage to the built and coastal environment (Byron Shire Council 2009b). Recent damage has been caused at two Byron Shire coastal areas: Belongil Beach just west of the main town of Byron Bay, and New Brighton, 30 km to the north. Belongil Beach has urbanised rapidly since the late 1980s, and by 2007 there were approximately 180 houses with an estimated value of \$1–5 million each (McDonald 2007).

Byron Shire's planned retreat policy was introduced in 1988 when the council was under administration by the state government. The planned retreat policy 'enables the temporary occupation of lands subject to coastline hazards, until such time that the risk to the development from coastal processes is unacceptable thus requiring the relocation or removal of development from that property' (Byron Shire Council 2009a). The policy pertinent to this case study states that dwellings need to be set back 20 m (or 50 m where applicable) from the erosion zone – if the erosion zone moves landwards then the owner is required to relocate or remove the dwelling (Byron Shire Council 2009a). Information on coastal hazards has been provided to landowners since the introduction of the policy, to inform prospective buyers of restrictions. This has been done via Section 149 Certificates, which are planning documents issued by council to provide information about property zoning, relevant state, regional and local planning controls, and any other property affectations. Byron's planned retreat policy was mostly latent until May 2009 when erosion from storms across the region directly threatened two coastal areas, triggering implementation of the policy. The implementation stimulated significant debate in the local community and a considerable level of attention from local and national news media.

Research approach and methods

This case study investigates the community discourse that occurred in the local media, specifically in letters to the editor. We describe the various dimensions of uncertainty evident in community response to Byron Bay's local council's planned retreat policy. News media have considerable power in a community through their ability to inform and shape public opinion (Holliman 2004; Wahl-Jorgensen 2005). In particular, local newspapers (a media outlet targeted to a specific location) play a major role in opinion formation within a community about local issues (Besley *et al.* 2008) and cover different aspects of natural hazards from those covered by national newspapers (Rashid 2011). At the local scale, newspapers help reveal the local understanding of the phenomena, of evidence/knowledge and associated issues, different expectations of authority and action, and normative judgements of values, concerns, risk and uncertainty.

Letters to the editor in particular tend to highlight community values (Wilkins 1993) as they use the language and themes of the people who are stimulated enough to write and submit them (Wahl-Jorgensen 2001). They use a communicative space to enable people to express their daily concerns and opinions and interact with their community (Rojas-Lizana 2007). This type of data reveals how stories and language can provide an opportunity to gain a rich picture of the breadth and depth of various perspectives (Sandercock 2004) that are not necessarily available through more formal means (Innes 1998). They are not intended to be a representative sample of views but more a portrayal of the perceptions, understandings, misunderstandings and misinformation that can occur in a community conflict. The letters are filtered through the newspaper's editorial process, which tends to favour the contextually rich stories of individuals (Wahl-Jorgensen 2001) rather than 'prepared' letters by more organised interests (Reader 2008).

The cyclic, episodic and self-reinforcing nature of news is well understood (Vasterman 2005; Crawley 2007). This case study discusses a news cycle relating to coastal erosion following a storm event at Belongil Beach in May 2009, that triggered Byron Shire Council's policy of planned retreat in the coastal erosion zone. The letters to the editor (LTE) in two local newspapers were examined from 1 June 2009 to 31 December 2010 inclusive. Byron Shire supports two local newspapers: the *Byron Shire Echo*, an independently owned newspaper established in 1986 as a 'provocative community-based newspaper' (Martin and Ellis 2002, p. 1) which has a circulation of 21 000, and the *Byron Shire News* which is part of the APN News and Media Ltd and has a circulation of 17 500. The data set included 114 letters which were entered into a data management package QSR Nvivo 8 (Nvivo). A coding scheme was developed using an iterative process which involved developing an initial list of coding categories from the first four months of data. These categories were adjusted during analysis of the letters collected over the subsequent 14 months. In this chapter we discuss only the themes related to uncertainty.

Dimensions of uncertainty in community responses to local planning for sea level rise

Scientific uncertainty

Community discourse that highlighted scientific uncertainty about sea level rise within the local context of Byron Bay focused on a local coastal erosion event; erosion, as several authors have pointed out, is only going to get worse with climate change. Local residents were particularly interested in exposing uncertainty about the 'facts' of the cause of the coastal erosion experienced at Belongil. Analysis of LTE suggested that the cause of the erosion at that beach is attributed by many to a rock wall constructed and maintained by Byron Shire Council, approximately 500 m from the main erosion site. Byron Shire Council assembled evidence from its coastal planning process to argue that the rock wall has no impact on the erosion affecting houses at Belongil. Interestingly, LTE authors drew on the same published reports to argue the opposite case:

> Council has on record several reports warning them that the rock wall in front of the swimming pool will cause massive erosion at Belongil.

The conflict over the 'facts' or science is significant because it provides the evidence base for legal action or compensation claims if responsibility can be clearly attributed to Byron

Shire Council, or directly or indirectly to the NSW government. Such conflict over the science, particularly deflection of responsibility for the cause of hazards, is common in environmental contexts (Bickerstaff and Walker 2002; Harrison *et al.* 1996). In the UK, Milligan and O'Riordan (2007) reported attempts to shift the responsibility or culture of blame for coastal erosion to off-shore activities (in this case, aggregate extraction). These were held responsible by those affected around north Norfolk.

The contest over the interpretation and representation of scientific knowledge undermines the credibility of a local council. Some LTE authors suggested that Byron Shire Council was taking a stance based on its politics and ideology – as 'a posse of green fascists' who 'all live well clear of the affected areas' – and was only utilising the science that supported its own position:

> It is of course very convenient for Council to uphold the view that damage from the rock walls only extends to 200 metres along the beach as this absolves them from responsibility for damage further along.

The credibility of the council was further undermined by its seemingly rationalistic planning process because, as one letter said, 'A lot of people don't care about facts'. Therefore the council stance was judged inadequate in considering the diversity of knowledge and values and ignoring equity aspects such as financial and social costs to residents. As one author entreated:

> Put aside the politics, the experts' reports, the petty jealousies or whatever else may come into the equation – and consider this simple fact – we live here. Our homes are here, our families are here, we live here! What part of that don't they understand?

Epistemological uncertainty

Uncertainty over the cause of the erosion also drew out a range of knowledge available within the Byron Bay community: several LTE authors offered observations based on direct experience or historical documents. For example, one resident wrote:

> I used to walk from the Wreck to beyond Belongil nearly every day. I observed that the sand cliffs created after storm activity, even though they are only loosely packed sand, stay vertical for a surprisingly long time and only started to rebuild once the top edge had crumbled and formed a mound at the base of the sand cliff.

Historical local knowledge sometimes plays a role in the choice of property purchase[1] because of buyers' perception of the risk and uncertainty of the coastline:

> In the early 80s I could have bought a very cheap house at Belongil, but being aware of the damage done by the 74 cyclone, decided that risking the possible loss of the value of my principal asset would be a pretty stupid thing to do. Not to mention the sleepless nights listening to the roaring ocean and waiting to end up in it. It was a no-brainer, the sensible thing to do was to 'live well clear of the affected areas'.

Local knowledge contributed a number of new ideas, potential solutions which were debated or supported by others. These included suggestions such as council resuming and leasing back property, soft engineering options such as 'reinstalling the ancient slope of the dune', beach nourishment or dune revegetation, hard structures on the coastal zone such as artificial reefs, or even just learning to live with the erosion.

Some authors expressed annoyance that their local knowledge – which sometimes included daily observations over decades – was not taken into account in local council decisions. This caused frustration, especially when local experience and observations conflicted with the official science-based narrative and ultimately undermined the credibility of council's processes. For example:

> Been to the beach lately? Hello, this is a letter from the Real World … Head down to Belongil, and have a look at the beach that is not supposed to be there. Actually have a damn good look … So it appears the real world, in which things actually happen, is again at variance with a plan.

In considering whose knowledge was included or privileged in the discourse, it is clear that different actors had varying capacity for engagement. The property owners of the erosion-affected beaches of Belongil and New Brighton were organised and represented as residents' associations. However, it was the Belongil residents whose voice dominated the discourse because they had high levels of resources with which to influence the decision-making processes. This meant that the knowledge and values of Belongil residents became privileged in the discussion because the group was more organised and powerful (and more litigious) than residents affected in other areas (albeit less dramatically, and in less iconic and affluent coastal areas of the shire). The portrayal of Belongil residents was also a source of conflict aired through the discourse. Those with little sympathy for their situation described them as 'rich bastards', 'King Canutes' or a 'bunch of whingeing millionaires' who should have recognised the risks of their property purchase and to whom the principle of *caveat emptor* (buyer beware) should apply. This contrasted with how Belongil property owners were described by one of their own, as 'homeowners have worked very hard all their lives to be able to dream and choose where they would like to live and retire'.

Ethical uncertainty

The issue of ethical uncertainty was a strong one in the discourse, with equity concerns being expressed over the rights of the individual versus the community and the ecosystem. Also important was the equity of costs.

There was a high level of conflict over property rights versus the public asset of the beach, described as 'the veranda of the continent, where Australians are at their egalitarian best'. While many argued for the beach as an important community asset, others favoured protection of individual assets:

> Well it's about the beach and that it should be available to any one at any time. How many tens of thousands of miles of beach about Australia do these selfish people actually need? Forget about the community that lives behind it, forget about the infrastructure and all the energy that went into creating that.

Significant concern was expressed at the potential loss of the beach for its ecological importance, particularly as a significant nesting site for shore birds. One LTE author asked:

> Why does a green council set out to destroy the Little Tern, a wetland, and a 6000-year-old dune in favour of a man-made feature and man-made erosion?

The views and values of non-coastal residents were considered less relevant or even gratuitous. This also featured in the discourse on whose opinions counted in discussions of coastal erosion, i.e. who was 'local' and so allowed to contribute to local community discussion. For example, a boundary between 'insider' and 'outsider' knowledge was maintained in the statement by one Belongil resident:

> And why would he care anyway? He lives in Mullumbimby!

All residents in the region would be affected, directly or indirectly, through council funds being spent on erosion works or on compensation for property owners. They would also be affected if private landowners were permitted to defend their properties, resulting in the erosion of the public asset of the beach. Concern was expressed over the financial cost to the community, particularly with the council spending 'large amounts of our money on solicitors and barristers to stop residents protecting their property'.

Uncertainty of scales of responsibility

Given the important role of state governments in determining local council powers, roles and responsibilities it is surprising that there is little mention of state climate change policy and its implication in the Byron Bay context. The NSW government introduced a draft NSW Sea Level Rise Policy Statement in November 2009, followed by *NSW Coastal Planning Guidelines: Adapting to Sea Level Rise* in August 2010. Yet these significant steps in state planning policy were barely mentioned in LTE. The local authority was clearly recognised as the lead agency in this discourse and the one responsible for developing and implementing the planned retreat policy. The highly local-centric nature of this discourse offered by council was interesting, given the hybrid nature of local councils as representing local constituents and as implementers of coastal policy on behalf of the state government.

Discussion

Localness in the discourse published in local media was framed in terms of who was a local and thus allowed to be involved, how scientific evidence was treated, whose knowledge was included and who was responsible for coordinating action. The implications of this localness were significant for the local government as it attempted to balance the expectations of a changing community with the policy requirements of state government. This study demonstrates that the uncertainties around a complex issue go beyond a lack of scientific understanding, and invoke various social perspectives operating at the local scale. Rather than scientific evidence providing a credible basis for planning, scientific modelling and expert opinion was refuted and contested through local debate. Local council efforts need to balance expert and local knowledge contributions using a more sophisticated approach, beyond the usual commissioning of expert reports followed by regulatory planning based on short periods of public comments and consultative committees. The study also raises questions over who should be involved in planning processes and how power differentials need to be managed through such processes.

Ethical as well as epistemological and scientific dimensions of uncertainty can fuel conflict. As media analysis of this case study shows, key issues concerning residents were focused on issues of fairness (managing risk against sea level impact) and responsibility (for bearing the cost of sea level impact). Mechanisms to expose and manage conflict are also required, to reduce division between different groups within a community and prevent long-lasting damage. In short, to be resilient, local council policies and community interactions need to be based on the assumption that coastline values and communities will be undergoing constant change, with inherent uncertainties.

Experience in the UK (Tompkins *et al.* 2008) and New Zealand (Turbott and Stewart 2006) shows that thoughtful, prolonged and participatory processes such as scenario planning can help to reduce community conflict over a receding coastline and can build trust in local government. The need for more creative engagement is imperative for issues that are not salient for a community until a threshold is reached.

Conclusion

Discourses surrounding a community response to local government policy highlight the fact that, while climate change is inherently a global issue, the impacts will be experienced most acutely at the local level. The localness articulated in Byron Bay resident discourse provides a useful illustration of how a global issue is experienced and constructed at the local scale. Byron Bay residents reflect the experience of many coastal communities that struggle to manage a dynamic coastal zone in the best interests of all while dealing with local pressure and scrutiny and with global broad uncertainties posed by sea level rise (Poulter *et al.* 2009; Moser 2005). The evidence-base, epistemological and scalar dimensions of uncertainty have important implications for the design of resilient local institutions. Councils have to actively handle community responses and their government obligations to manage the complex impacts and responses to sea level rise – and more broadly, climate change – in a credible way.

The analysis of the community discourse presented in this chapter highlights what planning theorists (Forester 1982; Fischer 2000) describe as the 'real world' complexity of planning, in which the possibilities of policy action depend on the context in which they are developed. We consider that understanding such locally constructed community responses can result in more effective coastal policies to support adaptation to sea level rise.

Endnote

1 On a methodological note, the rationality of people who considered coastal property and decided not to buy based on the perceived risk to the property is an important viewpoint to capture. It emerges through a data collection method such as LTE but is often difficult to elicit through other means.

References

Aarup T, Church J and Wilson SD (Eds) (2010) *Understanding Sea-level Rise and Variability.* Wiley-Blackwell, Chichester.

Abel N, Gorddard R, Harman B, Leitch A, Langridge J, Ryan A and Heyenga S (2011) Sea level rise, coastal development and planned retreat: analytical framework, governance principles and an Australian case study. *Environmental Science and Policy* **14**, 279–288.

ABS (2006) Basic community profile, Byron Shire. Cat. No. 2001.0. 2006 Community Profile Series. Australian Bureau of Statistics, Canberra.

Alexander KS, Ryan A and Measham TG (in press) Managed retreat of coastal communities: understanding responses to projected sea level rise through social functionalist frameworks. *Journal of Environmental Planning and Management* forthcoming.

Allmendinger P, Barker A and Stead S (2002) Delivering integrated coastal-zone management through land-use planning. *Planning Practice and Research* **17**, 175–196.

Anon (2010) Caravan park plans draw in tents response. ABC local radio. <http://www.abc.net.au/news/stories/2010/09/16/3013672.htm?site=northcoast&section=news> (accessed 01.10.10).

Australian Government (2010) Adapting to climate change in Australia: an Australian government paper. Australian Government, Canberra.

Berwick M (2007) The challenge of coastal governance. In *Federalism and Regionalism in Australia: New Approaches, New Institutions?* (Eds AJ Brown and JA Bellamy). ANU E-Press, Canberra.

Besley JC, McComas KA and Trumbo CW (2008) Local newspaper coverage of health authority fairness during cancer cluster investigations. *Science Communication* **29**, 498–521.

Bickerstaff K and Walker G (2002) Risk, responsibility and blame: an analysis of vocabularies of motive in air-pollution(ing) discourses. *Environment and Planning A* **34**, 2175–2192.

Binning C, Young M and Cripps E (1999) *Beyond Roads, Rates and Rubbish: Opportunities for Local Government to Conserve Native Vegetation.* Environment Australia, Canberra.

Boykoff MT (2008a) Lost in translation? United States television news coverage of anthropogenic climate change, 1995–2004. *Climatic Change* **86**, 1–11.

Boykoff MT (2008b) The cultural politics of climate change discourse in UK tabloids. *Political Geography* **27**, 549–569.

Bray M, Hooke J and Carter D (1997) Planning for sea-level rise on the South Coast of England: advising the decision-makers. *Transactions of the Institute of British Geographers* **22**, 13–30.

Brooks N, Adger WN and Kelly PM (2005) The determinants of vulnerability and adaptive capacity at the national level and the implications for adaptation: adaptation to climate change: perspectives across scales. *Global Environmental Change Part A* **15**, 151–163.

Bulkeley H and Betsill M (2005) Rethinking sustainable cities: multilevel governance and the 'urban' politics of climate change. *Environmental Politics* **14**, 42–63.

Burnley IH and Murphy PA (2004) *Sea Change: Movement from Metropolitan to Arcadian Australia.* University of New South Wales Press, Sydney.

Byron Shire Council (2009a) Draft Byron Shire Coastal Zone Management Plan. Plan prepared by Byron Shire Council, Mullumbimby.

Byron Shire Council (2009b) History of coastal erosion and coastal planning in Byron Shire. Report prepared by Byron Shire Council, Mullumbimby.

Cooper JAG and McKenna J (2008) Social justice in coastal erosion management: the temporal and spatial dimensions. *Geoforum* **39**, 294–306.

Cooper JAG, Anfuso G and Del Rio L (2009) Bad beach management: European perspectives. *Geological Society of America* Special paper **460**, 167–179.

Crawley CE (2007) Localized debates of agricultural biotechnology in community newspapers: a quantitative content analysis of media frames and sources. *Science Communication* **28**, 314–346.

Cutter SL and Emrich CT (2006) Moral hazard, social catastrophe: the changing face of vulnerability along the hurricane coasts. *Annals of the American Academy of Political and Social Science* **604**, 102–112.

DECC&W (2009) NSW Sea Level Rise Policy Statement. NSW Government, Sydney.

Dept of Climate Change (2009) Climate change risks to Australia's coasts: a first pass national assessment. Dept of Climate Change, Canberra.

Dollery B (2005) The state of play in Australian local government. Working Paper Series, 5-2005.

Doulton H and Brown K (2009) Ten years to prevent catastrophe? Discourses of climate change and international development in the UK press. *Global Environmental Change* **19**, 191–202.

England P (2006) Climate change: what are local governments liable for? Urban Research Program, Griffith University, Brisbane.

Failing L, Gregory R and Harstone M (2007) Integrating science and local knowledge in environmental risk management: a decision-focused approach. *Ecological Economics* **64**, 47–60.

Fischer F (2000) *Citizens, Experts and the Environment: The Politics of Local Knowledge.* Duke University Press, Durham.

Flyvberg B (2001) *Making Social Science Matter: Why Social Inquiry Fails and How it Can Succeed Again.* Cambridge University Press, Cambridge.

Forester J (1982) Understanding planning practice: an empirical, practical and normative account. *Journal of Planning Education and Research* **2**, 59–71.

Forester J (2007) No longer muddling through: institutional norms fostering dialogue, getting the facts and encouraging mediated negotiations. In *Institutions and Planning.* (Ed N Verma) pp. 37–60. Elsevier, Oxford.

Friedmann J (1973) *Retracking America: A Theory of Transactive Planning.* Anchor Press, New York.

Gurran N, Squires C and Blakeley E (2005) *Meeting the Sea Change Challenge: Sea Change Communities in Coastal Australia.* National Sea Change Task Force. Planning Research Centre, University of Sydney, Sydney.

Harrison CM, Burgess J and Filius P (1996) Rationalizing environmental responsibilities: a comparison of lay publics in the UK and the Netherlands. *Global Environmental Change Part A* **6**, 215–234.

Harvey N and Caton B (2003) *Coastal Management in Australia.* Oxford University Press, Oxford.

Harvey N and Clark B (2007) Policy implications for Australian coastal communities affected by sea-level rise. *Just Policy: A Journal of Australian Social Policy* **46**, 52–59.

Harvey N and Woodroffe C (2008) Australian approaches to coastal vulnerability assessment. *Sustainability Science* **3**, 67–87.

Harvey N, Clouston E and Carvalho P (1999) Improving coastal vulnerability assessment methodologies for integrated coastal zone management: an approach from South Australia. *Australian Geographical Studies* **37**, 50–69.

Haward M (1996) Institutional framework for Australian ocean and coastal management. *Ocean and Coastal Management* **33**, 19–39.

Hayward B (2008) 'Nowhere far from the sea': political challenges of coastal adaptation to climate change in New Zealand. *Political Science* **60**, 47–59.

Healey P (1999) Institutionalist analysis, communicative planning, and shaping places. *Journal of Planning Education and Research* **19**, 111–121.

Healey P (2006) *Collaborative Planning: Shaping Places in Fragmented Societies.* 2nd edn. Macmillan, London.

Holliman R (2004) Media coverage of cloning: a study of media content, production and reception. *Public Understanding of Science* **13**, 107–130.

Hunter J (2009) Estimating sea-level extremes under conditions of uncertain sea-level rise. *Climatic Change* **99**, 331–350.

Innes JE (1996) Planning through consensus building. *Journal of the American Planning Association* **62**, 460–473.

Innes JE (1998) Information in communicative planning. *Journal of the American Planning Association* **64**, 52–63.

Innes JE and Booher DE (2003) The impact of collaborative planning on government capacity. In *Annual Conference of the Association of Collegiate Schools of Planning*, Baltimore.

IPCC (2007) Climate change 2007: impacts, adaptation and vulnerability. Contribution of Working Group II to the 4th Assessment Report of the Intergovernmental Panel on Climate Change.

Johnstone B (2010) Beachfront price eroded at auction. *Sydney Morning Herald*, 30.10.10. <smh.domain.com.au/real-estate-news/beachfront-price-eroded-at-auction-20101029-1771k.html>.

Lahsen M (2007) Trust through participation? Problems of knowledge in climate decision-making. In *The Social Construction of Climate Change* (Ed M Pettenger) pp. 173–196. Ashgate Publishing, Aldershot.

Lane MB and McDonald G (2005) Community-based environmental planning: operational dilemmas, planning principles and possible remedies. *Journal of Environmental Planning and Management* **48**, 709–731.

Leitch AM, Harman B and Lane MB (2010) From blueprint to footprint: climate change and the challenge for planning. In *Adaptation to Climate Change: Law and Policy*. (Eds T Bonyhady, A Macintosh and J McDonald) pp. 63–79. Federation Press, Sydney.

Lipman Z and Stokes R (2003) Shifting sands: the implications of climate change and a changing coastline for private interests and public authorities in relation to waterfront land. *Environmental and Planning Law Journal* **406**, 1–18.

Lowe JA and Gregory JM (2010) A sea of uncertainty. *Nature Reports Climate Change* **1004**, 42–43.

Lyne P (2010) Save Seal Rocks. *Focus Mag*. <http://focusmag.com.au/mgl/featured/save-seal-rocks> (accessed 01.10.10).

Martin F and Ellis R (2002) Dropping in, not out: the evolution of the alternative press in Byron Shire 1970–2001. *Transformations* **2**, 1–15.

McDonald J (2007) The adaptation imperative: managing the legal risks of climate change. In *Climate Law in Australia* (Ed T Bonyhady) pp. 124–142. Federation Press, Sydney.

McDonald J and England P (2007) A risky climate for decision making: the legal liability of development authorities for climate change impacts. In *Proceedings of QELA 2007Conference: Your System or Mine?* Kingscliff, 16–18 May 2007.

Measham TG, Preston BL, Smith TF, Brooke C, Gorddard R, Withycombe G and Morrison C (2011) Adapting to climate change through local municipal planning: barriers and challenges. *Mitigation and Adaptation Strategies for Global Change* **16**, 889–909.

Miles RL, Kinnear S, Marshal C, O'Dea G and Greer L (2009) Assessing the socio-economic implications of climate change (coral bleaching) in the Great Barrier Reef catchment: synthesis report. Report to the Marine and Tropical Sciences Research Facility. Cairns Reef and Rainforest Research Centre Ltd.

Milligan J and O'Riordan T (2007) Governance for sustainable coastal futures. *Coastal Management* **35**, 499–509.

Morrison C, Lusher B and Withycombe G (2009) What's the magic number: how can councils consider and plan for rising sea levels? *New South Wales Coastal Conference*. Ballina, NSW.

Moser S (2000) Appendix F. Community responses to coastal erosion: implications of potential policy changes to the National Flood Insurance Program. In *Evaluation of Erosion Hazards*. H John Heinz II Center for Science, Economics and the Environment. Project prepared for the Federal Emergency Management Agency, Washington.

Moser SC (2005) Impact assessments and policy responses to sea-level rise in three US states: an exploration of human-dimension uncertainties. *Global Environmental Change Part A* **15**, 353–369.
Moser SC and Dilling L (2004) Making climate hot: communicating the urgency and challenge of global climate change. *Environment* **46**, 32–46.
Nicholls N (1999) Cognitive illusions, heuristics and climate prediction. *Bulletin of the American Meteorological Society* **80**, 1385–1397.
Norman B (2009) Principles for an intergovernmental agreement for coastal planning and climate change in Australia. *Habitat International* **33**, 293–299.
O'Hagan AM and Ballinger RC (2010) Implementing integrated coastal zone management in a national policy vacuum: local case studies from Ireland. *Ocean and Coastal Management* **53**, 750–759.
Oreskes N (2004) The scientific consensus on climate change. *Science* **306**, 1686.
O'Riordan T and Ward R (1997) Building trust in shoreline management: creating participatory consultation in shoreline management plans. *Land Use Policy* **14**, 257–276.
Poulter B, Feldman RL, Brinson MM, Horton BP, Orbach MK, Pearsall SH, Reyes E, Riggs SR and Whitehead JC (2009) Sea-level rise research and dialogue in North Carolina: creating windows for policy change. *Ocean and Coastal Management* **52**, 147–153.
PWD (1978) Byron Bay/Hastings Point erosion study. Report No. PWD 78026. Coastal Engineering Branch, NSW Public Works Dept.
Rashid H (2011) Interpreting flood disasters and flood hazard perceptions from newspaper discourse: tale of two floods in the Red River valley, Manitoba, Canada. *Applied Geography* **31**, 35–45.
Reader B (2008) Turf wars? Rhetorical struggle over 'prepared' letters to the editor. *Journalism* **9**, 606–623.
Rojas-Lizana I (2007) Letters to the editor in the local press: genre and discourse in everyday texts from Australia and Chile. PhD thesis. University of Queensland, Brisbane.
Ryan A, Gorddard R, Abel N, Leitch AM, Alexander K and Wise R (2011) Perceptions of sea level rise risk and the assessment of managed retreat policy: results from an exploratory community survey in Australia. Climate Adaptation National Research Flagship. CSIRO, Canberra.
Rydin Y (2003) *Conflict, Consensus and Rationality in Environmental Planning: An Institutional Discourse Approach.* Oxford University Press, Oxford.
Sandercock L (2004) Towards a planning imagination for the 21st century. *Journal of the American Planning Association* **70**, 133–141.
Sarewitz D (2004) How science makes environmental controversies worse. *Environmental Science and Policy* **7**, 385–403.
Tol RSJ, Klein RJT and Nicholls RJ (2008) Towards successful adaptation to sea-level rise along Europe's coasts. *Journal of Coastal Research* **24**, 432–442.
Tompkins EL and Adger NW (2005) Defining response capacity to enhance climate change policy: mitigation and adaptation strategies for climate change. *Environmental Science and Policy* **8**, 562–571.
Tompkins EL, Few R and Brown K (2008) Scenario-based stakeholder engagement: incorporating stakeholders' preferences into coastal planning for climate change. *Journal of Environmental Management* **88**, 1580–1592.
Turbott C and Stewart A (2006) Managed retreat from coastal hazards: options for implementation. Environment Waikato Technical Report 2006, pp. 48–89. Environment Waikato, Hamilton.

van Kerkhoff L and Lebel L (2006) Linking knowledge and action for sustainable development. *Annual Review of Environment and Resources* **31**, 445–477.

Vasterman PLM (2005) Media-hype: self-reinforcing news waves, journalistic standards and the construction of social problems. *European Journal of Communication* **20**, 508–530.

Wahl-Jorgensen K (2001) Letters to the editor as a forum for public deliberation: modes of publicity and democratic debate. *Critical Studies in Media Communication* **18**, 303–320.

Wahl-Jorgensen K (2005) Threats to local journalism in Britain: the market vs the right to communicate. *Javnost-The Public* **12**, 79–93.

Walsh K, Cai W, Hennessy K, Jones R, McInnes K, Nguyen K, Page C and Whetton P (2004) Climate change in Queensland under enhanced greenhouse conditions. Final report 1997–2002. Report to Queensland government. CSIRO Atmospheric Research, Brisbane.

Watson PJ (2011) Is there evidence yet of acceleration in mean sea level rise around mainland Australia? *Journal of Coastal Research* **27**, 368–377.

Weingart P, Engels A and Pansegrau P (2000) Risks of communication: discourses on climate change in science, politics and the mass media. *Public Understanding of Science* **9**, 261–283.

Wild River S (2002) The environmental implications of the local–state antinomy in Australia. PhD thesis. Australian National University, Canberra.

Wild River S (2006) Australian local government attempts to deliver beneficial environmental outcomes. *Local Environment* **11**, 719–732.

Wilkins L (1993) Between facts and values: print media coverage of the greenhouse effect 1987–1990. *Public Understanding of Science* **2**, 71–84.

Wynne B (1989) Sheep farming after Chernobyl: a case study in communicating scientific information. *Environment* **31**, 10–15, 33–9.

Wynne B (1992) Uncertainty and environmental learning: reconceiving science and policy in the preventative paradigm. *Global Environmental Change* **2**, 111–127.

Wynne B (2002) Risk and environment as legitimatory discourses of technology: reflexivity inside out? *Current Sociology* **50**, 459–477.

10 RISK AND CLIMATE CHANGE IN BRAZILIAN COASTAL CITIES

Leila da Costa Ferreira, Rafael D'Almeida Martins, Fabiana Barbi, Alberto Matenhauer Urbinatti, Fernanda Oliveira de Souza, Thales Haddad Novaes de Andrade and Leonardo Freire de Mello

Introduction

Climate change has become one of the most pressing issues in the 21st century (Giddens 2009; Hogan 2009a). Since 1988, the Intergovernmental Panel on Climate Change (IPCC), established by the United Nations and the World Meteorological Organization, has been engaged in understanding the functioning of the global climate system. The greatest scientific advances of this movement came with the most recent reports (3rd and 4th Assessment Reports), which expressed relative consensus on global warming and concluded with higher confidence that climate change is unequivocal and the increase in average temperature of the Earth's surface results from human activities combined with natural variation of the global climate system (IPCC 2007). This represents an advance in the debate: previous positions and arguments persisted, defending, with some power of public persuasion, that the contemporary global warming had only natural causes resulting mainly from biogeological cycles of the planet. According to such positions, being a natural problem, there would be nothing to do about climate change apart from accepting it and coping with it (Giddens 2009).

Although the scientific community recognises that the Earth's temperature is subject to long-term variation, the argument is that it is possible to assess the contribution of human activities to the 'expanded greenhouse effect' and to observe that the pace of this phenomenon has been verified since the advent and internationalisation of the industrial revolution and the western life-style, in more or less accelerated rhythms in several countries (Wilbanks and Kates 1999; Beck 2000; Pittock 2005). The current and apparently unprecedented changes challenge the adaptive and responsive capacity of the planet since they are still largely unknown (Young *et al.* 2006), despite several scientific initiatives that seek to provide a better understanding of these processes on an integrated planetary scale (Steffen *et al.* 2004; Rockström *et al.* 2009).

Living in a global society means facing many risks shaped by multiple stressors as there are great uncertainties in terms of what the world will be like in 30, 50 or 100 years. So far, it is only possible to mention probabilities and possible scenarios. Based on that, the highly modern society is involved with the challenge of managing increasing risks. In terms of responses to climate change, it has been argued that governments are important actors that play an important role in setting regulations, institutions and appropriate modes of governance for confronting these risks at different levels and scales (Giddens 2005, 2009; Bulkeley and Kern 2006; Alber and Kern 2008; Newell and Bulkeley 2010; Bulkeley 2010).

In Brazil, where major cities are located on coastal zones, the impacts of climate change are an extremely relevant issue. Coastal areas are considered particularly vulnerable to climate change (Adger 1999; Nicholls *et al.* 2007); at the same time they concentrate an expressive contingent of the population (McGranahan *et al.* 2007) and are development centres of the worldwide economy due to their infrastructure and natural resources (Satterthwaite *et al.* 2007). This chapter reflects on the challenges of climate change for these areas, how the issue could be framed and how local governments in Brazilian coastal cities in the state of São Paulo are prepared (or not) to this risk in terms of policy strategies and instruments (Fig. 10.1, colour section).

This debate is in its early stages in Brazil (TCU 2009; Martins and Ferreira 2009, 2010a, 2010b). This chapter represents the first attempt to investigate some of the political-institutional structures that can have some effect on relevant areas for intervention in terms of climate change at the local level in the 16 coastal cities of the state of São Paulo.

Risk and climate change

A reconsideration of the nature of current ecological crisis is suggested by Beck (1992, 1995, 2009) since the side effects of industrial production are characterised as a deep institutional crisis of the industrial society itself. This proposition can be applied to climate change, the causes of which are deeply rooted in modern societies – apparently innocuous and invisible gases such as methane (CH_4) and carbon dioxide (CO_2) are released in the atmosphere as development products (industrial agriculture, increased energy consumption and economic growth), altering its composition with incalculable consequences (Bulkeley 2001).

In the risk society, the technical-industrial development causes unpredictable, invisible, incalculable risks that require the society to reflect on the basis of social cohesion and examine its rationality foundations. The risks that characterise modern industrial society are closely related to the future, to the misfortunes actively assessed in relation to future possibilities and situations of which our society has little experience (Beck 1992, 2009; Giddens 2005, 2009). These characteristics are closely related to climate change risks and are influenced by the increasing process of globalisation (Young *et al.* 2006; Leichenko and O'Brien 2008).

It is hard to know and in many cases impossible to predict the magnitude of these risks before it is too late, since the society does not know what other changes will come nor the dangers they will bring. The challenge is how to establish effective policies to address different problems, since the future is uncertain and it is not possible to evaluate the risks with complete accuracy until the anticipated hazards materialise (Giddens 2009). The required political action and intervention at the international, national and local levels will have a decisive effect in setting boundaries to avoid dangerous global warming and support adaptation efforts that are increasingly underway (Parry *et al.* 2008; Parry 2009), particularly in the developed countries (Moser 2009 – US; Hulme and Neufeldt 2010 – Europe).

Climate change and local governments

Global changes in climate, environment, economy, societies, governments, institutions and cultures converge in localities. The effects of these changes at the local level in turn contribute to global changes and are affected by them (Leichenko and O'Brien 2008). As a result, connecting the local and global scales across a broad range of disciplines and problems – integrated assessments of population, economy, technology and environmental changes – enables a potentially deeper understanding of global environmental change, including climate change, in all its complexity (Wilbanks and Kates 1999; Adger *et al.* 2009).

According to data from several countries, local governments may control policies that can deal with about 30–50% of national greenhouse gas emissions (Robinson and Gore 2005). Local governments are usually responsible for local transportation, development planning and energy management (Coenen and Menkveld 2002; Bulkeley and Betsill 2003). Since a large number of people live and work within city boundaries and are under local government jurisdiction, these governments are in a privileged position to influence many activities that generate large greenhouse gas emissions (Yienger *et al.* 2002).

With urbanisation, a centralised state is less necessary since an urban population can solve most of its problems locally. This reality has led the most developed countries to adopt a state structure with broad participation of local governments. Part of the literature that analyses sustainability frameworks (Castells 1983; Cahn 1995; Buttel 2000) clearly indicates the importance of decentralisation, i.e. the participation of several social actors and stakeholders in finding solutions to environmental problems, at the local level where opportunities are more noticeable.

In this sense, local government has an important role in responding to climate change risks (Satterthwaite *et al.* 2007). Currently, many local governments in developing countries have low institutional and response capacity to deal with the problems that affect the quality of life, especially the poorest with less coping capacity and weak ties in social protection networks (Satterthwaite *et al.* 2007). However, climate change does not reach only the poor. Its impacts affect different sectors of society, the public sphere and the market (Beck 1992, 2009; Wilbanks *et al.* 2007; Satterthwaite *et al.* 2007; Satterthwaite 2008; Bartlett *et al.* 2009). Local governments are considered critical in addressing many of the causes and consequences of climate change as they are close to the locations of relevant activities and events (Wilbanks and Kates 1999; Robinson and Gore 2005; Satterthwaite *et al.* 2007; Puppim de Oliveira 2009; Bulkeley 2010).

Table 10.1 summarises some local and state government responsibilities in areas related to climate change and risk management that need to be components of a comprehensive adaptation strategy to climate change. Local governments are usually responsible for (Satterthwaite *et al.* 2007; Satterthwaite 2008):

- finance – financial and accounting management of municipal budget, rates collection and management, licences and fees;
- engineering and public construction – construction and maintenance of the public space;
- urban/local development – land use regulation, urban zoning, real estate registration and urban planning;
- health and public hygiene – collection, distribution and treatment of drinking water, pollution control, collection and treatment of solid waste, sanitary hygiene, public area cleaning, medical and ambulatory services;
- social urban policies – housing, schools, daycare centres, youth, the elderly etc.;
- civil defence and emergency response – disaster relief, fire brigades, ambulance services and rescue;
- public administration and human resources – various administrative tasks and responsibilities including human resources management.

Climate change in Brazilian coastal cities

Coast of São Paulo state

In Brazil, many important cities and infrastructure are located along more than 8600 km of coastline, making the impacts of climate change extremely relevant for the present and future

Table 10.1: Role of local governments in aeas with effects on climate change

Local government role	Long-term planning	Prevention of disasters	Response to extreme events	Reconstruction
Built environment				
Construction code	High	n.a.	High	High
Land use regulation and real estate registration	High	Low	n.a.	High
Maintenance and construction of building and public space	High	Low	n.a.	High
Urban planning and zoning	High	n.a.	High	High
Infrastructure				
Water treatment	High	Low	High	High
Wastewater treatment	High	Low	High	High
Drainage system	High	Low	High	High
Streets, avenues, bridges	High	n.a.	High	High
Electricity	High	Low	High	High
Landfill	High	Low	n.a.	High
Services				
Protection against fire	High	Low	High	Low
Public order, police	Average	High	High	Low
Solid waste collection	High	High	High	High
Education	Average	Average	n.a.	n.a.
Health and public hygiene	Average	Average	High	High
Public transportation	Average	High	High	High
Social welfare	Average	High	High	High
Responses to disasters and civil defence	n.a.	n.a.	High	High

Source: Adapted from Satterthwaite (2008, p. 26).

of these areas (Nicolodi and Petermann 2010). It is estimated that more than 20% of the Brazilian population live on coastal cities (Zamboni and Nicolodi 2008). São Paulo, the largest Brazilian state in terms of population concentration and economic development, has many of its infrastructures located on its coastal region. The ports of Santos and São Sebastião, respectively the biggest port in South America and the main distribution centre for Brazilian inland oil, along with the industrial district of Cubatão and the tourism industry are examples of the importance of this region for the social and economic development of the country (Martins and Ferreira 2010c).

The coast of São Paulo extends for about 700 km. The state includes 16 cities with over 2 million inhabitants (Table 10.2) and has diversified ecosystems with estuaries, mangroves, sandbanks, caves, rivers, rocky shores and about 150 islands (SMA 2005). There are several protected areas within its territory as it contains a great proportion of Brazil's Atlantic Forest remnants, one the three most threatened ecosystems on the planet and a hotspot for biodiversity (Aidar *et al.* 2001). Although there are a number of measures to protect these important ecosystems, they have been endangered by increasing tourism activities, modern transportation infrastructures, the oil and gas industry and illegal exploitation of natural resources (Martins and Ferreira 2010c).

Table 10.2: Population of the coastal cities of São Paulo state

City	Population
Cananéia	13 562
Iguape	29 321
Ilha Comprida	9536
Cubatão	124 284
Guarujá	312 504
Itanhaém	91 788
Mongaguá	45 376
Peruíbe	55 743
Praia Grande	249 266
Santos	432 213
São Vicente	331 581
Bertioga	44 341
Caraguatatuba	91 397
Ilhabela	27 690
São Sebastião	71 290
Ubatuba	84 137
Total	**2 014 029**

Source: IBGE (2009).

In terms of regional and socio-economic aspects, the northern part of the coast of São Paulo state is being targeted by a series of development interventions and expressive investments in infrastructure, especially in the area with the cities of Caraguatatuba and São Sebastião (Hogan 2009b; Martins and Ferreira 2010c). The construction of a number of new facilities is expected, including:

- a road bypass connecting the cities of São Sebastião and Caraguatatuba;
- a regional landfill to minimise the chaotic situation of solid waste in the region;
- a prison facility in Caraguatatuba;
- a natural gas treatment plant (already in an advanced stage of construction).

Despite the absence of consolidated data, it is expected that this region will be affected by climate change as well as by the ongoing interventions and projects, the results of which might lead to a second phase of disruptive urban growth and expansion with consequences that could be dramatic for the integrity of the Atlantic Forest and for different social groups, especially the people who live in the area (Marengo 2007; SMA 2009; Hogan 2009b; Martins and Ferreira 2010c). It is appropriate to consider how climate change can affect this area and the links between the processes. The chapter starts by mapping public agencies, municipal bodies and specific institutions whose main focus is to act on environmental issues, in order to assess the existence of institutional capacity that could be mobilised in the future to address climate change.

Political-institutional strategies

Regional-level strategies

In Brazil, the city of São Paulo was a pioneer in approving a specific law to deal with climate change. The municipal law (no. 14933), approved in 2009, established the reduction of 30% in total greenhouse gas emissions of the city (Martins and Ferreira 2010d). It enforced public transportation priority, gradual reduction in fossil fuel consumption and dependence, a compulsory recycling program and reduction of solid waste together with energy efficiency criteria and eco-building.

Some months later, the state of São Paulo approved a similar law (State Law no. 13.798, November 2009), becoming the second Brazilian state to approve a type of regulation (the first was the state of Amazonas in 2007) which presents tangible mitigation targets. The Climate Change State Policy expects a reduction of at least 20% in total CO_2 emissions by 2020 in relation to the 2005 level. Every five years, until 2020, the state government will be able to set intermediate targets to reach the 20% goal (Lucon and Goldemberg 2010).

Despite being considered substantial advances in addressing the climate change problem, these laws rely on a network of political-institutional structures and governance arrangements for effective implementation (see Table 10.1). Thus, it is fundamental to investigate the presence of some of these structures in the 16 coastal cities of São Paulo state as those cities do not present any clear climate change response in terms of specific laws or policy frameworks, unlike the city and state of São Paulo. In order to assess the institutional and administrative capacity of these coastal cities to address the climate change issue, a mapping exercise was carried out. It focused on the presence of governmental bodies and institutions interlinked with sectors and policies with some level of effect and jurisdiction to address climate change.

Local political-institutional structures

Our mapping exercise focused on the executive and legislative municipal bodies as well as the presence of secretariats, departments, other public agencies and city council commissions that had some history and capability in dealing with areas that can affect or be affected by climate change. As the coastal cities did not have particular institutions to respond to climate change, looking at municipal realities was a strategy to frame the climate change issue within the current administrative structures (Ferreira 1998; Martins and Ferreira 2010c). The research was not exhaustive and the presence of structures and institutions that work with different sectors guarantees neither effectiveness nor an adequate response to climate change, but it is a starting point for understanding the capacity of São Paulo coastal cities in terms of their ability to formulate and implement climate strategies.

Table 10.3 shows that many of the investigated cities had political-institutional structures that could be mobilised to address the climate change issue, especially in terms of specific secretariats, governmental agencies and departments. Our survey highlighted the presence of structures dedicated to social, economic and urban development policies, as well as environment and health.

Most of the city councils had commissions which would be able to intervene and govern climate change by authority (Bulkeley *et al.* 2009), basically through legislation such as the laws approved in the city and state of São Paulo, that aim at mitigating greenhouse gas emissions, creating and mobilising the necessary conditions and resources for adaptation strategies.

The area of planning and management did not possess specific municipal structure in some of the analysed cities, which might cause problems for coordinating intersectoral and multi-dimensional actions (Bulkeley *et al.* 2009; Martins and Ferreira 2010b). In the case of transport, only five cities directly managed the public transport of their localities. In the other cities it was indirectly administrated through public-private consortiums and municipal companies. In many cities, the transport sector was one of the main sources of greenhouse gas emission; thus, enabling ways to govern this sector is fundamental to mitigation actions (Collier 1997; Robinson and Gore 2005; Bulkeley *et al.* 2009; Martins and Ferreira 2010b). Only the city of Ilha Comprida did not possess a specific structure to deal with environmental

Table 10.3: Political-institutional structure in coastal cities of São Paulo state

Area of action*	Cities	No. of cities
Economic development		
Secretariats, authorities (Agriculture, Fishery, Industry, Commerce, Science, Technology)	Cananéia, Caraguatatuba, Cubatão, Guarujá, Ilha Comprida, Itanhaém, Mongaguá, Peruíbe, Santos, São Sebastião, São Vicente, Ubatuba	12
Social development		
Secretariats, authorities (Social Action, Social Promotion, Social Assistance, Citizenship)	Bertioga, Cananéia, Caraguatatuba, Cubatão, Guarujá, Ilhabela, Ilha Comprida, Itanhaém, Mongaguá, Peruíbe, Praia Grande, Santos, São Vicente, Ubatuba	14
Urban development		
Secretariats, authorities (Housing, Public Construction, Public Services, Infrastructure)	Bertioga, Cananéia, Caraguatatuba, Cubatão, Guarujá, Iguape, Ilhabela, Ilha Comprida, Itanhaém, Mongaguá, Peruíbe, Praia Grande, Santos, São Vicente, São Sebastião, Ubatuba	16
Environment		
Secretariats, authorities	Bertioga, Cananéia, Caraguatatuba, Cubatão, Guarujá, Iguape, Ilha Bela, Itanhaém, Mongaguá, Peruíbe, Praia Grande, Santos, São Sebastião, São Vicente, Ubatuba	15
Planning (management)		
Secretariats, authorities	Caraguatatuba, Cubatão, Guarujá, Ilha Comprida, Itanhaém, Mongaguá, Peruíbe, Santos, São Vicente, Ubatuba	10
Health		
Secretariats, authorities	Bertioga, Cananéia, Caraguatatuba, Cubatão, Guarujá, Iguape, Ilha Bela, Ilha Comprida, Itanhaém, Mongaguá, Peruíbe, Praia Grande, Santos, São Sebastião, São Vicente, Ubatuba	16
Security		
Secretariats, authorities, directories (Public Security, Urban Security, Social Defence, Civil Defence)	Caraguatatuba, Guarujá, Itanhaém, Mongaguá, Peruíbe, Santos, São Sebastião, São Vicente, Ubatuba	9
Transport		
Secretariats, authorities	Caraguatatuba, Iguape, Itanhaém, Mongaguá, São Vicente	5

* Some secretariats, authorities or directories have a specific area of intervention (e.g. Planning, Environment) while others act in more than one area at the same time (e.g. Planning and Environment).

policies at the local level, although such activities could be incorporated into its urban development department.

These data may suggest that the studied local governments have taken the first steps in internalising the climate change issue. It is important to have these structures and institutions, but the implementation of actions thus far has not been sufficient to face the challenging scenarios projected by the scientific community (IPCC 2007; Parry *et al.* 2008; Parry 2009).

Local strategies on climate change

The implementation of local policies regarding climate change requires the integration of several areas (Table 10.1) and may stand against many challenges posed by traditional political practices and institutions (Puppim de Oliveira 2009; Bulkeley *et al.* 2009; Martins and Ferreira 2010c, 2010d; Bulkeley 2010). Pre-existing patterns of political intervention sometimes inhibit assessment and the formation of complex interactions between the sectors involved in climate change policies (Murgida and Natenzon 2009).

In the coastal cities of São Paulo state, local authorities have argued that many problems in dealing with such issues arise from lack of resources as well as advanced political instruments and alternative interactions between offices and departments (Bulkeley *et al.* 2009; Martins and Ferreira 2010d). The sectoral and social interests at stake and the existing political capacities provide only a narrow space for long-term and far-reaching climate policy implementation.

Science is providing increasingly reliable data, methodologies, tools and frameworks to help society adapt to the impacts of climate change and variability, but local governments on the coast of São Paulo state are still struggling to provide the most basic urban services (e.g. adequate housing, sanitation, health care) for a proportion of their population (Satterthwaite *et al.* 2007; Martins and Ferreira 2010d). The lack of financial and human resources (Martins and Ferreira 2010c, 2010d) combined with the high level of uncertainty about climate change impacts, the magnitude of risks (Beck 1992, 2009; Giddens 2009) and the need for climate change scientific data at the local level also contribute to the lack of consistent action in this region on climate change risks (Martins and Ferreira 2010d).

The involvement of specific offices and departments with civil society organisations is another aspect that claims attention. Some coastal cities have strong social movements and organisations that focus on environmental issues such as fishing, forest conservation and recovery. In some cities, the participation of such groups in the urban governance is quite expressive in confronting many initiatives that they consider damaging. However, due to the lack of legal frameworks or adequate policy coordination, there are no comprehensive policies linking these groups to the local authorities on climate change issues, although some seminars and public hearings have been organised. These engagements remain only occasional and voluntary opportunities. The local management councils (deliberative public spaces) are not currently very active, but they represent an important arena where policies could be designed and evaluated with the participation of several stakeholders and policy-makers (Martins and Ferreira 2010d).

Even though the coastal cities do not have climate specific strategies, the fact that a broader environmental agenda is institutionalised through some type of legislation not only at the regional level (Lucon and Goldemberg 2010) but also in most of the cities (Table 10.4) may suggest possible ways for the internalisation of climate change issues in local governance. The political-institutional structure is favourable, however, it is necessary to consider the coordination of different strategies cutting across existing policies that are related to climate change to some extent, and levels of governance (local, state and national).

Conclusion

Based on discussions within the social sciences (e.g. Beck 1992, 2009), the risks arising from climate change are characterised by intangibility – the results of the risks only become evident when the hazards and other effects materialise. They can be altered, deepened, dramatised or minimised as they are open to social definition and construction. The risks are closely related to the future and to the misfortunes actively assessed in relation to future possibilities.

Table 10.4: Environmental legislation in the coastal cities of São Paulo state

Cities	Municipal organic law	Land use planning	Environmental code	Urban development planning	Regional ecological-economic zoning	Conservation units Acts	Others
Bertioga		Yes	Yes				
Cananéia							Yes
Cubatão	Yes						
Guarujá	Yes	Yes					
Ilha Comprida	Yes						
Itanhaém	Yes	Yes		Yes			
Peruíbe	Yes						
Praia Grande		Yes		Yes	Yes		
Santos	Yes	Yes				Yes	Yes
São Vicente		Yes					Yes
São Sebastião							Yes
Ubatuba	Yes						

Political action and intervention at international, national and local levels will have a decisive effect on establishing limits to global warming and setting the bases for adaptation to ongoing and unavoidable climate change (Parry *et al.* 2008; Giddens 2009). International agreements are essential and international organisations, NGOs, private sector and community-based organisations all have vital roles to play. However, the state in its all levels of government is key to avoiding dangerous climate change and implementing the necessary adaptation measures.

In terms of specific climate change policies, the debate in Brazil is still in the early stages and there are a number of issues for the coming years. Despite climate change being a new subject in the political and policy arena, laws approved since 2007 at the national, state and municipal levels not only aim at mitigating greenhouse gas emissions but also coordinate adaptation efforts by combining risk management and disaster risk reduction.

Local governments, not only in Brazil but elsewhere, are essential in addressing many of the causes and consequences of climate change. They are close to where these activities and events take place, considering areas such as the built environment, infrastructure and public services, which have direct and indirect effects on climate change (Table 10.1).

This chapter was the first attempt to investigate the political-institutional structures that have some effect on climate change at the local level in the 16 coastal cities of São Paulo state. Analyses (Tables 10.3, 10.4) showed that these cities have secretariats, departments, directories and other types of governmental bodies as well as environmental legislation that can be mobilised for climate change actions.

This chapter argued that the existing structures, and the alternatives proposed until now by those institutions, are far from sufficient to respond effectively to the magnitude and complexity of the climate change problem. As our debates and decisions will determine the future quality of life on Earth – not only for humans but also for diverse ecosystems – there are a number of ethical issues and dilemmas to be raised and addressed. There are reasonable arguments to defend the social engagement in collective and political dialogues, negotiation and proposals that could enhance the response capacity to challenges so that responses would not be only palliative. It means moving towards reformist views that do not address the root causes of global warming and society's vulnerability that are expressed by the social injustice and the unsustainable patterns of human development.

This chapter has contributed towards understanding the capacity of some São Paulo coastal cities to deal with climate change, building upon existing political-institutional structures. As it was the first step in this direction, the chapter has raised more questions than it was able to answer. How do local and regional actors react on climate change issues? To what extent do the local political actors have a complete view of the climate change problem and the project impacts that are expected to happen or are already under way? What makes some cities better prepared to respond to the challenges than others? These and many other questions should be addressed in future studies.

References

Adger WN (1999) Social vulnerability to climate change and extremes in coastal Vietnam. *World Development* **27**, 249–269.

Adger WN, Eakin H and Winkels A (2009) Nested and teleconnected vulnerabilities to environmental change. *Frontiers in Ecology and the Environment* **7**, 150–157.

Aidar MPM, Godoy JRL, Bergmann J and Joly CA (2001) Atlantic forest succession over calcareous soil, Parque Estadual Turístico do Alto Ribeira – PETAR, SP. *Revista Brasileira de Botânica* **24**, 455–469.

Alber G and Kern K (2008) Governing climate change in cities: modes of urban climate governance in multi-level systems. In *OECD International Conference: Competitive Cities and Climate Change*. 2nd Annual Meeting of the OECD Roundtable Strategy for Urban Development, 9–10 October, Milan.

Bartlett S, Dodman D, Hardoy J, Satterthwaite D and Tacoli C (2009) Social aspects of climate change in urban areas in low- and middle-income countries. In *World Bank Urban Research Symposium*, Marseilles.

Beck U (1992) *Risk Society: Towards a New Modernity*. Sage, London.

Beck U (1995) *Ecological Politics in an Age of Risk*. Cambridge University Press, Cambridge.

Beck U (2000) *What is Globalization?* Polity Press, Cambridge.

Beck U (2009) *World at Risk*. Polity Press, Cambridge.

Betsill M and Bulkeley H (2007) Looking back and thinking ahead: a decade of cities and climate change research. *Local Governments* **12**, 447–456.

Bulkeley H (2001) Governing climate change: the politics of risk society? *Transactions of the Institute of British Geographers* **26**, 430–447.

Bulkeley H (2010) Cities and the governing of climate change. *Annual Review of Environment and Resources* **35**, 229–253.

Bulkeley H and Betsill MM (2003) *Cities and Climate Change: Urban Sustainability and Global Environmental Governance*. Routledge, London.

Bulkeley H and Kern K (2006) Local government and the governing of climate change in Germany and UK. *Urban Studies* **43**, 2237–2259.

Bulkeley H, Schroeder H, Janda K, Zhao J, Armstrong A, Chu SY and Ghosh S (2009) Cities and climate change: the role of institutions, governance and urban planning. In *World Bank Urban Research Symposium*, Marseilles.

Buttel FH (2000) Sociologia ambiental, qualidade ambiental e qualidade de vida: algumas observações teóricas. In *Qualidade de vida e riscos ambientais*. (Eds S Herculano, MFS Porto and CM Freitas). Niterói, EDUFF.

Cahn MA (1995) *Environmental Deceptions: The Tension between Liberalism and Environmental Policy-making in the United States*. State University of New York, New York.

Castells M (1983) *The City and the Grassroots: A Cross-cultural Theory of Urban Social Movements*. University of California Press, Los Angeles.

Coenen FM and Menkveld M (2002) The role of local authorities in a transition towards a climate-neutral society. In *Global Warming and Social Innovation: The Challenge of a Climate-neutral Society*. (Eds MTJ Kok, WJW Vermeulen, APC Faaij and D de Jager) pp. 107–125. Earthscan Publications, London.

Collier U (1997) Local authorities and climate protection in the European Union: putting subsidiary into practice? *Local Environment* **2**, 39–57.

Ferreira LC (1998) *Sustentabilidade e Políticas Públicas no Brasil*. Ed. Boitempo, São Paulo.

Giddens A (2005) *Mundo em descontrole. O que a globalização está fazendo de nós*. Record, Rio de Janeiro.

Giddens A (2009) *The Politics of Climate Change*. Polity Press, Cambridge.

Hogan DJ (2009a) População e mudanças ambientais globais. In *População e mudança climática: dimensões humanas das mudanças ambientais globais*. (Eds DJ Hogan and E Marandola Jr). Ed. Unicamp, Campinas.

Hogan DJ (Coord.) (2009b) Urban growth, vulnerability and adaptation: social and ecological dimensions of climate change on the coast of São Paulo. Research project approved by the Fapesp Global Climate Change Program.

Hulme M and Neufeldt H (2010) *Making Climate Change Work for Us: European Perspectives on Adaptation and Mitigation Strategies*. Cambridge University Press, Cambridge.
IBGE (Instituto Brasileiro de Geografia e Estatística) (2009) Brasil: População. <http://www.ibge.gov.br/paisesat/main.php> (accessed 02.10.10).
IPCC (Intergovernmental Panel on Climate Change) (2007) *Climate Change 2007: Impacts, Adaptation and Vulnerability*. Contribution of Working Group II to the 4th Assessment Report of the Intergovernmental Panel on Climate Change. (Eds ML Parry, OF Canziani, JP Palutikof, PJ van der Linden and CE Hanson). Cambridge University Press, Cambridge.
Leichenko RM and O'Brien KL (2008) *Environmental Change and Globalization: Double Exposures*. Oxford University Press, New York.
Lucon O and Goldemberg J (2010) São Paulo – the 'other' Brazil: different pathways on climate change for state and federal governments. *Journal of Environment and Development* **19**, 335–357.
Marengo JA (2007) *Caracterização do clima no Século XX e Cenários no Brasil e na América do Sul do Século XXI derivados dos Modelos de Clima do IPCC*. Ministério do Meio Ambiente, Brasília.
Martins RDA and Ferreira LC (2009) Assessing the research on the human dimensions of global environmental change in Latin America. *Teoria e Pesquisa* **18**, 31–52.
Martins RDA and Ferreira LC (2010a) The research on the human dimensions of global environmental change in Latin America: looking back, moving forward. *International Journal of Climate Change Strategies and Management* **2**, 264–280.
Martins RDA and Ferreira LC (2010b) Governing climate change at the city level: tales from two global cities in Brazil. In *Conference on the Human Dimensions of Global Environmental Change, Social Dimensions of Environmental Change and Governance*. Freie Universität Berlin, 8–9 October, Berlin.
Martins RDA and Ferreira LC (2010c) Double exposure on the northern coast of the state of São Paulo, Brazil. In *Conference on the Human Dimensions of Global Environmental Change, Social Dimensions of Environmental Change and Governance*. Freie Universität Berlin, 8–9 October, Berlin.
Martins RDA and Ferreira LC (2010d) Enabling climate change in urban areas: a local governance approach. *INTERthesis* **7**, 241–275.
McGranahan G, Balk D and Anderson B (2007) The rising tide: assessing the risks of climate change and human settlements in low elevation coastal zones. *Environment and Urbanization* **19**, 17–37.
Moser SC (2009) *Good Morning America! The Explosive Awakening of the US to Adaptation*. NOAA and California Energy Commission, Charleston and Sacramento.
Murgida and Natenzon CE (2009) Social downscaling: a few reflections on adaptation in urban environments. In *Public Policy Mitigation and Adaptation to Climate Change in South America*. (Eds PL Dias, WC Ribeiro, JLS Neto and J Zuillo Jr). IEA/USP, São Paulo.
Newell P and Bulkeley H (2010) *Governing Climate Change*. Routledge, New York.
Nicholls RJ, Wong PP, Burket VR, Codignotto JO, Hay JE, McLean RF, Ragoonaden S and Woodroffe CD (2007) Coastal systems and low-lying areas. In *Climate Change 2007: Impacts, Adaptation and Vulnerability*. Contribution of Working Group II to the 4th Assessment Report of the Intergovernmental Panel on Climate Change. (Eds ML Parry, OF Canziani, JP Palutikof, PJ van der Linden and CE Hanson). Cambridge University Press, Cambridge.
Nicolodi JL and Petermann RM (2010) Mudanças climáticas e a vulnerabilidade da zona costeira do Brasil: aspectos ambientais, sociais e tecnológicos. *Revista da Gestão Costeira Integrada* **10**, 151–177.

Parry ML (2009) Climate change is a development issue, and only sustainable development can confront the challenge. *Climate and Development* **1**, 5–9.
Parry ML, Palutikof J, Hanson C and Lowe J (2008) Squaring up to reality. *Nature Reports Climate Change* **2**, 1–3.
Pittock AB (2005) *Climate Change: Turning up the Heat*. Earthscan Publications, London.
Puppim de Oliveira JA (2009) The implementation of climate change related policies at the subnational level: an analysis of three countries. *Habitat International* **33**, 253–259.
Robinson PJ and Gore CD (2005) Barriers to Canadian municipal response to climate change. *Canadian Journal of Urban Research* **14**, 102–120.
Rockström J, Steffen W, Noone K, Persson Å, Chapin FS III, Lambin E, Lenton TM, Scheffer M, Folke C, Schellnhuber H, Nykvist B, de Wit CA, Hughes T, van der Leeuw S, Rodhe H, Sörlin S, Snyder PK, Costanza R, Svedin U, Falkenmark M, Karlberg L, Corell RW, Fabry VJ, Hansen J, Walker B, Liverman D, Richardson K, Crutzen P and Foley J (2009) A safe operating space for humanity. *Nature* **461**, 472–475.
Satterthwaite D (2008) *Climate Change and Urbanization: Effects and Implications for Urban Governance*. UN Expert Group Meeting on Population Distribution, Urbanization, Internal Migration and Development. UN/POP/EGM-URB/2008/16. New York.
Satterthwaite D, Huq S, Pelling M, Reid H and Lankao PR (2007) Adapting to climate change in urban areas: the possibilities and constraints in low- and middle-income nations. Discussion Paper No.1. International Institute for Environment and Development, London.
SMA (Secretaria de Meio Ambiente) (2005) *Planejamento Ambiental. Zoneamento Ecológico-Econômico – Litoral Norte*. SMA/CPLA, São Paulo.
SMA (Secretaria de Meio Ambiente) (2009) *Litoral Norte. Metodologia para Avaliação Ambiental Integrada de Projetos*. SMA/CPLA, São Paulo.
Steffen WL, Sanderson A, Tyson PD, Jäger J, Matson PA, Moore B III, Oldfield F, Richardson K, Schellnhuber HJ, Turner BL II and Wasson RJ (2004) *Global Change and the Earth System: A Planet under Pressure*. Springer-Verlag, Berlin.
TCU (Tribunal de Contas da União) (2009) Auditorias de natureza operacional sobre políticas públicas e mudanças climáticas: adaptação das zonas costeiras brasileiras. TCU, Brasilia.
Wilbanks T and Kates RW (1999) Global change in local places: how scales matter. *Climatic Change* **43**, 601–628.
Wilbanks T, Lankao PR, Bao M, Berkhout F, Cairncross S, Ceron J-P, Kapshe M, Muir-Wood R and Zapata-Marti R (2007) Industry, settlements and society. In *Climate Change 2007: Impacts, Adaptation and Vulnerability*. Contribution of Working Group II to the 4th Assessment Report of the Intergovernmental Panel on Climate Change. (Eds ML Parry, OF Canziani, JP Palutikof, PJ van der Linden and CE Hanson). Cambridge University Press, Cambridge.
Yienger J, Brown L and Skinner N (2002) *Experiences of ICLEI's Cities for Climate Protection Campaign (CCP): A Focus on Asia*. Proceedings of IGES/APN Mega-City Project. Institute for Global Environmental Strategies.
Young O, Berkhout F, Gallopin GC, Janssen MA, Ostrom E and van der Leeuw S (2006) The globalization of socio-ecological systems: an agenda for scientific research. *Global Environmental Change* **16**, 304–316.
Zamboni A and Nicolodi JL (2008) *Macrodiagnóstico da Zona Costeira e Marinha do Brasil*. Ministério do Meio Ambiente, Secretaria de Mudanças Climáticas e Qualidade Ambiental, Brasília.

11 VULNERABILITY ANALYSIS, RISK AND DELIBERATION: THE SYDNEY CLIMATE CHANGE ADAPTATION INITIATIVE

Tom Measham and Benjamin L. Preston

Introduction

The premise of this chapter extends from the case put forward by Ferreira *et al.* (Ch. 10, this volume) that climate change, and the way society responds to it, are examples of Beck's (1992) risk society thesis. In essence, invisible and dispersed risks such as climate change represent the side effects of modernity. Carbon dioxide, which is difficult to see and track, is released as a by-product of production and consumption. The impacts of these by-products are distant through time and space, and understanding the risks associated with them depends on specialist scientific knowledge (Bulkeley 2001). Science becomes drawn into the political process as we have seen through the Intergovernmental Panel on Climate Change assessment reports (IPCC 2007), the Stern Review (Stern 2007) and the Garnaut (2008) report in Australia. An important question for science emerging from Beck's (1992) thesis concerns the extent to which we can make the intangible risks of modernity more visible (Preston *et al.* 2011; Shaw *et al.* 2009).

Beck (1992) described the transition to risk society in terms of reflexivity, a process of self-confrontation and attempting to comprehend and legitimise the risks of modernity. This process is complicated by scientific uncertainty. So we see in climate change considerable concern over precise forecasts for impacts and a politics of stalling for further information (Bulkeley 2001). At one extreme, society may find itself knowing exactly what it should have done but finding it is too late anyway. Conversely, an increasing concern focuses on the risk of failing to act based on available information, for fear of legal repercussions down the track. Social scientists concerned with climate change have argued that the degree of certainty required to respond to climate change varies according to context. It has been argued that reasonable and defensible decisions to address climate change impacts can occur based on partial certainty (Dessai and Hulme 2007). These examples help to break the deadlock of seeking complete and comprehensive information prior to climate action.

The risk society thesis claims that failure to produce a coordinated, coherent and comprehensive response to the intrinsic risks of modernity leads inevitably to an institutional crisis, represented by a division of mainstream politics and an emerging subpolitics. Herein lies a key challenge in the risk society thesis – how to address complex risks in an integrated way as they are drawn out of the political sphere and into the insulated domains of expert committees, business strategies and political lobbying. In this chapter, we consider the risk society thesis in relation to local governments in the Sydney region. In part, our focus on local government stems from a pragmatic issue, namely that a peak body, the Sydney Coastal Council Group (SCCG) which supports these municipalities and lobbies on their behalf, engaged a team of researchers to collaborate with a view towards facilitating climate adaptation action at the local scale.

In addition to our pragmatic rationale for focusing on local government, we argue that this layer of governance helps to address the problems of the risk society thesis. Beck's general theory of governance has little to say about the municipal scale but we focus on it because this is where policy meets the public in a tangible way. We note that local governments in Australia (and elsewhere) have no constitutional basis. Rather, they are the delegated agents of a higher power such as a state in Australia (Wild River 2006), or a province in the case of Canada (Ivey *et al.* 2004). These local entities are concerned with the details of domestic reality which higher levels of government are unwilling or unable to address, such as the collection of waste, provision of childcare, maintenance of public parks and, particularly important for climate change, vulnerability – the siting of residential dwellings and businesses (Pini *et al.* 2007). Despite this subordinate role (or perhaps because of it), some authors argue that local governments often represent a ground for grassroots action and innovation, where leadership and fresh thinking towards environmental sustainability can emerge outside formal leadership roles (Brown 1997, 2005).

In what has been described by theorists as local–state antimony, rather than seeing themselves as the servants of state bureaucracies, local government representatives hold the counter position that they serve local communities and in many cases agitate against state politics which conflict with local values (Wild River 2005). We argue that this antinomy represents a ripe context in which to explore risk governance of complex environmental challenges. The local scale is particularly relevant in the case of climate change, where an invisible, dispersed problem presents tangible local risks such as sea level rise and storm events (Ivey *et al.* 2004). However, we need to expand our theoretical palette beyond Beck's risk society (1992) to recognise that the state is itself a complicated suite of different institutions, which can compete, collaborate or operate in ignorance of each other across a range of social and environmental challenges and concepts of risk (Renn and Schweizer 2009; Renn *et al.* 2011). Furthermore, there is value in taking a participatory approach to vulnerability assessment, in order to involve stakeholders in working through the technical complexities of risks and to consider potential implications for those who live with the assessment outcomes (Renn 2004; Preston *et al.* 2011).

Case study: the Sydney Climate Change Adaptation Initiative

The Sydney Climate Change Adaptation Initiative was conducted from 2006 to 2008 to assist coastal municipalities in the Sydney basin to understand and respond to climate risks. The project was developed as a collaboration between the SCCG (a peak body for 15 municipalities) and the Commonwealth Scientific and Industrial Research Organisation (CSIRO). It is important to note that at the beginning of the project climate change denial was a dominant force in Australian politics; this altered during the course of the project, as reflected in a change of government at the federal level in 2007, partly due to increasing recognition of climate change. The national debates were echoed at the local and regional scales where, for example, storms and beach erosion were attributed to climate change. The case study area represented the coastal councils from Hornsby in the north to Sutherland in the south. The project team of 10 staff comprised professionals in environmental science, economics and human geography, and local government leaders and planning professionals. The project was conducted in three phases, which are described in the following sections.

Phase 1: Mapping vulnerability

The first phase of the project involved assessing and mapping relative vulnerability to climate impacts across the 15 participating council areas. This was initially conducted as a desktop exercise then presented to council representatives and staff for feedback, resulting in revisions to the assessment. The project adopted the IPCC (2001) definition of vulnerability, namely, 'the degree to which a system is susceptible to, or unable to cope with, adverse effects of climate change, including climate variability and extremes'. This definition incorporates the principle of coping capacity. Increasingly, there is recognition that the potential for adverse effects from a hazard such as climate change is not simply a function of the hazard itself, but also the ability of societies (e.g. through institutions, technologies and policies) to respond to that hazard (Ch. 2, this volume; Preston *et al.* 2008). In this project we approached vulnerability in terms of considering risk factors that represent susceptibility to harm. Five areas of potential climate impacts were considered for the vulnerability assessment:

- extreme heat and human health effects;
- sea level rise and coastal hazards;
- extreme rainfall and urban stormwater management;
- bushfire;
- natural ecosystems and assets.

Vulnerability was conceptualised as having three components: exposure, sensitivity and adaptive capacity (Preston *et al.* 2008). For example, calculating vulnerability in relation to heat stress involved indicators of exposure (e.g. number of days above 30°C), indicators of sensitivity (e.g. the proportion of the population over 65) and indicators of adaptive capacity (e.g. household income). The relationship between these different components is summarised in Figure 11.1 (colour section). A full list of indicators used to calculate extreme heat vulnerability is presented in Table 11.1.

It is important to clarify that the assessment process calculated relative vulnerability, namely, a ranking of how vulnerable each council was relative to the others (compared to some hypothetical benchmark of vulnerability). Hence, the vulnerability scores that were generated do not necessarily translate into an absolute probability of an adverse event or impact.

These indicators were integrated within a geographic information system to produce maps for each type of vulnerability by location, and a composite map which demonstrated net relative vulnerability across the study area (Fig. 11.2, colour section). The research team compared its assumptions underpinning the calculations with those of SCCG member council staff, who confirmed that they seemed reasonable. For some threats, such as sea level rise and bushfires, vulnerability maps generally agreed well with risk perceptions of council staff. For other threats, however, council staff often lacked sufficient knowledge to form an opinion on the appropriateness of the maps. This suggested that councils were far more aware of and sensitive to risks for which they had direct management authority or historical experience, or for which there were vocal community concerns. The results of this phase pointed to a number of hotspots within the Sydney region that were considered relatively more at-risk to the effects of climate change than other locations. They included north-western and southern Hornsby Shire Council, eastern Pittwater Council, the area between Sydney Harbour and Botany Bay (particularly Rockdale and Botany Bay City councils) and

Table 11.1: Vulnerability indicators for extreme heat and human health effects

Exposure indicators	Sensitivity indicators	Adaptive capacity indicators
1) Present average January maximum temperature (BOM stations)	1) % population ≥65 years of age[a]	1) % population completing year 12[a]
2) Present average January minimum temperature (BOM stations)	2) % population ≥65 years of age and living alone[a]	2) % population that speaks language other than English[a]
3) Present # days >30°C (BOM stations)	3) % population ≤4 years of age[a]	3) Median home loan repayment[a]
4) Projected change in average DJF maximum temperature in 2030 (25 km grid)	4) % of housing as multi-unit dwellings[a]	4) % home ownership[a]
5) Land cover (14 m grid)	5) Projected population growth to 2019[b]	5) Median household income[a]
6) Population density (census districts)		6) % households requiring financial assistance[a]
7) Road density (5 km grid)		7) % population with internet access[a] 8) Current ratios[c] 9) Per capita business rates[c] 10) Per capita residential rates[c] 11) Per capita community service expenses[c] 12) Per capita environment and health expenses[c]

a: Census districts.
b: Statistical local areas.
c: Local government areas.

northern Sutherland Shire Council (Preston *et al.* 2008). The causes of this vulnerability varied depending on local context.

This phase resulted in three key findings.

1 There was significant spatial variability across the SCCG region with respect to climate change vulnerability. The different classes of climate impacts varied from highly fragmented to concentrated in certain areas. This suggested the need to tailor responses to accommodate the unique challenges posed by different impacts across the area.
2 Demographic and socio-economic characteristics were equally relevant to biophysical hazards when determining the potential for harm.
3 The process of conducting the assessment was just as important as the outcome. Defining appropriate indicators of exposure, sensitivity and adaptive capacity and comparing them with council perspectives provided important insight and learning about the nature of vulnerability for the research team and project partners.

Phase 2: Community engagement workshops – vulnerability analysis as dialogue

Workshops were conducted with each of the 15 participating councils in the case study region to consider the results of the vulnerability mapping process and what they meant for each council. In addition to confronting the implications for each council area, the workshops focused on identifying opportunities and barriers for action. A total of 257 individuals took part across the 15 workshops. The workshops were attended by a broad cross-section of roles within local government including elected councillors representing local constituents and council staff such as social planners, land use planners, lawyers, engineers, senior managers, environmental officers and community engagement specialists (Smith *et al.* 2008a).

After reviewing the outputs from the vulnerability assessment, workshop participants created systems diagrams of key climate change drivers, impacts and management responses specific to their location. The systems diagrams were used to assist councils to identify their priority climate issues, which were then discussed in small groups in terms of barriers and opportunities for managing them. Many workshop participants were concerned about potential liabilities as a result of climate change and observed that current council plans and policies did not take climate change into account. Additionally, many participants were unclear as to the extent of local government statutory responsibilities in relation to climate change.

Following all the workshops, the research team grouped the broad range of concerns raised by participants into three cross-cutting issues. These were:

- community capacity and community expectations;
- infrastructure vulnerability and capacity, much of which is beyond the control of local governments;
- the nature of existing planning frameworks which limit council ability to respond to climate change.

These findings highlighted the precarious position of municipalities in the governance network, in that they exist at the interface of bottom-up, grassroots concerns of local communities as well as top-down constraints imposed by higher levels of the state.

Phase 3: Identifying barriers to action

To gain a deeper understanding of each of the cross-cutting issues that emerged from the workshops, a suite of three (sub) case studies was developed (Smith *et al.* 2008b). Three of the councils (Leichhardt, Mosman and Sutherland Shire) were selected for these more detailed studies. Thirty-three semi-structured interviews with representatives from those councils were conducted in April and May 2008. Participants held a range of roles within councils including elected councillors, senior managers, middle managers and operational staff. The interviews were recorded and transcribed verbatim. Interview responses were grouped into a series of themes coded in relation to:

- councils' current responsibilities for adapting to climate change;
- contextual, structural, procedural and outcomes considerations with regards to the three regional cross-cutting barriers (community, planning and infrastructure);
- preferred council climate change adaptation roles and responsibilities;
- what councils needed to do differently to achieve their climate change adaptation goals;
- respondents' expectations of this project.

In considering the responses, it was clear that the participating councils had already made significant progress in addressing climate change through greenhouse gas mitigation efforts. For some time, councils had been engaged in efforts to reduce emissions from council facilities and community constituents. This reflected widespread awareness of climate change across the participating councils and the growing momentum for substantive actions to reduce emissions. Such efforts did not, however, address the issue of adapting to the effects of climate change that cannot be avoided through mitigation. Local governments' more recent efforts on climate change adaptation illustrate the evolution of thinking and policy that manifests around emergent issues of public concern. Such efforts were tentative and ad hoc, composed of a mixture of community engagement and geotechnical risk assessment. Interviews with council staff and councillors provided a clear indication that, generally, the participating councils would like to exercise greater leadership in ensuring communities are appropriately prepared.

Participants acknowledged responsibility for a range of climate-related impacts including the need to revise details of stormwater run-off and the approach to coastal inundation. They also noted major barriers to climate adaptation, propagated from the New South Wales (NSW) and federal policy environment in which local government operates. At the time of the interviews, NSW state legislation and management frameworks relevant to local government activities assumed a stable climate. As a consequence, there was little ability for councils to manage climate risks within the set frameworks. This was exacerbated by the fact that other legislation placed restrictions on local government authority and decision-making with respect to building codes, rate increases and limits on growth and development. Collectively, these issues hampered progressive action by local government with respect to climate adaptation.

Further adaptation barriers stemmed from the organisational structures of local governments, resource availability and political priorities over decisions about climate risk in the context of other challenges faced by local government. Such barriers were a function of both exogenous and endogenous factors. On one hand, councils expressed limited capacity to cope with the broad range of regulatory and service demands imposed through NSW state legislation, given limited available resources including financial capital, technical information and expertise. On the other hand, it was also apparent that local government operations were structured around thematic 'silos', which compartmentalised expertise in core operational areas and limited internal deliberation and diffusion of knowledge. At the time of the interviews, some council sections that may well be vital to innovative approaches to managing climate risk didn't see it as part of their role. Our intention was to stimulate wider discussion of climate change vulnerability beyond the environmental section to other sections of council such as planners and engineers.

From reflexive modernisation to epistemological discourse

To address the intrinsic risks of modernity such as climate change, Beck (1992) proposed that society must go through a process of reflexive modernisation. Self-reflection and self-confrontation are crucial in this reflexive process to move beyond conflicts over accountability (Ch. 10, this volume). However, it is not clear from Beck's work where in the social system these processes can and should occur. In our case study, discussing the vulnerability maps with local governments directed their attention to local concerns and identified intraregional differences, thereby assisting councils to confront their own vulnerability. The purpose was to make the risk of climate impacts visible, even if not immediate (Giddens 2009). In some cases vulnerability was instantly visible from the maps, which clearly demonstrated some simple principles, for example that areas with less vegetation were less vulnerable to bushfire. The maps also helped to demonstrate cumulative effects, for example that areas with more elderly residents, high building density and a lack of planning instruments were more vulnerable to extreme heat events, due partly to increased exposure and partly to reduced capacity to address the issue. Given that one of the key elements of climate change is that the causes and impacts are distant in time and place, the mapping process at least helps to make the impacts more tangible at the local scale, even if the causes remain distant in time and place (Shaw *et al.* 2009; Beck 1992; Bulkeley 2001; Preston *et al.* 2011). At the same time, the vulnerability mapping caused council staff to reflect on their own knowledge and perceptions of risk, revealing gaps in understanding and confidence in management (Preston *et al.* 2009).

Although the first phase of the project was primarily a desktop assessment, the process of inviting feedback from council staff was crucial. Klinke and Renn (2002) argued that when risks are associated with uncertainty, scientific input is only the first step towards a wider process of deliberation. In practical terms, our approach was similar to that of Mukheibir and Ziervogel (2007), who encouraged the involvement of stakeholders in vulnerability assessment to increase social acceptance of the outcome. Preston *et al.* (2011) noted the importance of incorporating local knowledge and participation in both legitimising assessment results and facilitating learning. In theoretical terms, an inclusive approach to risk governance represents a normative assertion that integrating knowledge and values into risk assessment is best achieved by involving a wider set of actors whose respective knowledge leads to more effective, fair and morally acceptable outcomes (Renn and Schweizer 2009). This process led to the second phase of the study, which focused on interpreting the implications of the mapping phase and framing local climate impacts through multiple detailed workshops employing systems thinking methods (Ison *et al.* 2007).

The purpose of the workshops was to extend the deliberation that emerged from the mapping process. In this way, the workshops represented epistemological discourse, or deliberation among experts (Klinke and Renn 2006). Importantly, experts need not be scientists, but anyone with specialist knowledge who can contribute towards the best available understanding of the phenomenon in question. Thus in our study, council engineers and bushfire management professionals contributed to the vulnerability assessment process. The workshops sparked wide debate over how to respond to the particular climate risks faced in specific locations. More widely, the workshops cemented an acknowledgement that climate change is a real issue that needs local attention, rather than a vague concern over an ambiguous threat. This is not to say that the vulnerability assessments were completely unambiguous (Preston *et al.* 2009, 2011). Rather, the uncertainty contained within them was more familiar and could be related to the daily business of local government on a case by case basis. For example, the workshops sparked quite specific conversations among engineers on how much bigger drains should be in order to cope with larger floods, and the types of knowledge relevant to making such a determination. In addition, the inclusion of social and economic metrics in the vulnerability assessment expanded councils' traditional geotechnical framings of vulnerability. In this regard, the second phase of the study endorsed the social learning process proposed by Tàbara *et al.* (2010) who clearly demonstrated that 'more knowledge' alone is a flawed approach to climate adaptation – it depends on jointly defining and addressing problems.

To complement the vulnerability assessment (phase 1) and workshops to consider implications (phase 2), the focus of the third phase was to deepen understanding of the key issues emerging from the workshops, the existing council processes to manage them and the constraints which are limiting local adaptation to climate change in the case study region. The in-depth interviews and site visits in the third phase provided much more detail on specific actions conducted by councils and the barriers they face. It became clear, for example, that some councils had already taken responsibility for climate change mitigation. Their actions were mostly focused on energy and water efficiency and their initiatives preceded the project discussed here. In explaining their rationales, participants described a mix of ethical arguments and tactical reasons associated with political advantage gained through environmental leadership in particular communities. Above all, the interviews brought to light some aspects of risk management which were already within local government control, where councils could take more responsibility and develop innovative solutions.

The final phase of the project also highlighted the fact that councils were only starting to come to terms with incorporating adaptation into their scope of responsibilities. The project illustrated that, even with expanded awareness of adaptation concepts and opportunities, councils were heavily constrained in terms of their governance system. Two distinct sources of institutional limitations were evident in this study: those stemming from internal council architecture and those occurring at higher levels of government. The most acute internal limitation, noted across each of the three case study councils, was a strong tendency to assign climate adaptation (along with mitigation) to the environment section of the council. This derived from a legacy of thinking of climate change as an environmental issue. Institutional 'silos' are a historic issue, and climate adaptation is a renewed reason to improve integration within councils (Critchley and Scott 2005). The challenge for local government is to recognise climate adaptation as a cross-sectoral issue. This was starting to occur in some of the case study councils between the environment and water sections, with engineers considering the implications of flood and storm surge events. The second type of institutional limitation resulted from the multi-scale institutional context in which councils function (Amundsen *et al.* 2010). As Naess *et al.* (2005) demonstrated, councils frequently fulfil the role of implementing actions defined at higher tiers of government, with minimal room to manoeuvre. For example, it became clear that it is very difficult to take responsibility for climate impacts in planning decisions when the planning system does not recognise a changing climate (Measham *et al.* 2010).

We started this chapter with a discussion of Beck's (1992) risk society thesis and the need for reflexive modernisation in relation to climate change vulnerability in our Sydney case study. Considering the empirical material, it is hard to determine the extent to which reflexive modernisation occurred, nor indeed what reflexive modernisation means for local government in practical terms. Rather, we endorse the normative value of a more inclusive and deliberative approach to risk governance, in which the institutions within the state or their delegated subsidiaries are able to deliberate and adapt through the process of epistemological discourse (Renn and Schweizer 2009; Klinke and Renn 2006). Of course the outcome of this discourse is part of a complex set of interactions where cause and effect are difficult to determine with certainty. We can, however, draw attention to two developments in climate change impacts in the Sydney region that display hopeful signs. Subsequent to this research, the SCCG project partner lobbied the NSW government on behalf of the 15 participating councils to change the planning laws by which local governments operate, in order to recognise a changing climate. These efforts appear to have been successful – the planning system has been amended to incorporate the changes advocated by the SCCG and allow local governments to plan for sea level rise impacts (NSW 2010). This may not represent the sort of innovative institutional change implied by Beck's (1992) risk society, yet it does show that local government can challenge other elements of the state from the bottom up and push for change through processes such as epistemological discourse (Klinke and Renn 2006). These legal changes in the planning system have since been incorporated into Local Environmental Plans and applied in planning decisions on the siting of developments (Boydell 2010).

Conclusion

For a problem such as climate change which is distant in terms of its cause and effects, vulnerability analysis can play a crucial role in focusing discussion on what to do about the problem, rather than maintaining frustrating contests over accountability. To achieve this theoretically, we need to consider not only the theory of risk society but also the concept of inclusive risk governance. To achieve this practically, we need to make complex and distant risks more tangible and discuss their real-world implications. The project described in this chapter demonstrates that the visual nature of maps assisted councils to identify the local implications of a global challenge. This was an important step in reducing complexity to a scale which permitted discussion. This dialogue – through a series of facilitated workshops – focused attention on what to do about the risks of local climate impacts. Both steps represented vital elements of moving beyond (or at least through) conflicts over accountability through epistemological discourse. Finally, the process of clarifying the key underlying issues which are limiting climate adaptation, through detailed case analysis, helped bring to the surface the potential for innovative approaches to addressing climate impacts. Throughout these processes, the role of science was to enable discussion and debate rather than to prescribe solutions. The early signs emerging from this approach to managing local climate vulnerability have been encouraging.

Acknowledgements

This research was funded by the CSIRO Sustainable Regional Development Theme and the Climate Adaptation Flagship. Thanks to Russell Gorddard of CSIRO, Tim Smith of the University of the Sunshine Coast, Cassandra Brooke of WWF Australia and Geoff Withycombe, Beth Beveridge and Craig Morrison of the Sydney Coastal Councils Group for their invaluable contributions to the project upon which this chapter is based.

References

Amundsen H, Berglund F and Westskog H (2010) Overcoming barriers to climate change adaptation: a question of multilevel governance? *Environment and Planning C: Government and Policy* **28**, 276–289.

Beck U (1992) *Risk Society: Towards a New Modernity.* Sage, London.

Boydell ED (2010) Adaptation to climate change in practice: learning from a local government case study. Honours dissertation. Australian National University, Canberra.

Brown V (1997) *Managing for Local Sustainability: Policy, Problem-solving, Practice and Place.* Dept of Environment Sport and Territories, Canberra.

Brown VA (2005) Leadership in the local government sector: working from the inside out. In *The Natural Advantage of Nations: Business Opportunities, Innovation, and Governance in the 21st Century.* (Eds K Hargroves and MH Smith) pp. 289–295. Earthscan Publications, London.

Bulkeley H (2001) Governing climate change: the politics of risk society? *Transactions of the Institute of British Geographers* **26,** 430–447.

Critchley V and Scott J (2005) Changing governments: councils embracing the precautionary principle. In *Social Learning in Environmental Management: Towards a Sustainable Future.* (Eds M Keen, V Brown and R Dyball) pp. 146–169. James & James/Earthscan Publications, London.

Dessai S and Hulme M (2007) Assessing the robustness of adaptation decisions to climate change uncertainties: a case study on water resources management in the east of England. *Global Environmental Change* **17**, 59–72.

Ferreira LC, Martins RDA, Barbi F, Urbinatti AM, Souza FO, Andrade THN and Mello LF (2012) Risk and climate change in Brazilian coastal cities. In *Risk and Social Theory in Environmental Management.* (Eds T Measham and S Lockie) pp. 133–145. CSIRO Publishing, Melbourne.

Garnaut R (2008) *Garnaut Climate Change Review: Final Report.* Cambridge University Press, Melbourne.

Giddens A (2009) *The Politics of Climate Change.* Polity Press, Cambridge.

IPCC (Intergovernmental Panel on Climate Change) (2007) Summary for policymakers. In *Climate Change 2007: The Physical Science Basis.* Contribution of Working Group I to the 4th Assessment Report of the Intergovernmental Panel on Climate Change. (Eds ML Parry, OF Canziani, JP Palutikof, PJ van der Linden and CE Hanson). Cambridge University Press, Cambridge.

IPCC (Intergovernmental Panel on Climate Change) (2001) *Climate Change 2001: The Scientific Basis.* Contribution of Working Group I to the 3rd Assessment Report of the Intergovernmental Panel on Climate Change. Cambridge University Press, Cambridge.

Ison R, Blackmore C, Collins K and Furniss P (2007) Systemic environmental decision making: designing learning systems. *Kybernetes* **36**, 1340–1361.

Ivey JL, Smithers J, De Loë RC and Kreutzwiser RD (2004). Community capacity for adaptation to climate-induced water shortages: linking institutional complexity and local actors. *Environmental Management* **33**, 36–47.

Klinke A and Renn O (2002) A new approach to risk evaluation and management: risk-based, precaution-based and discourse-based strategies. *Risk Analysis* **22**, 1071–1094.

Klinke A and Renn O (2006) Systemic risks as challenge for policy making in risk governance. *Forum: Qualitative Social Research*, **7**, article 33 [online].

Measham TG, Preston BL, Smith TF, Brooke C, Gorddard R, Withycombe G and Morrison C (2011) Adapting to climate change through local municipal planning: barriers and challenges. *Mitigation and Adaptation Strategies for Global Change* **16**, 889–909.

Mukheibir P and Ziervogel G (2007) Developing a municipal adaptation plan (MAP) for climate change: the city of Cape Town. *Environment and Urbanization* **19**, 143–158.

Murphy (2012) Managing risk under uncertainty. In *Risk and Social Theory in Environmental Management.* (Eds T Measham and S Lockie) pp. 17–26. CSIRO Publishing, Melbourne.

Naess LO, Bang G, Eriksen S and Vevatne J (2005) Institutional adaptation to climate change: flood responses at the municipal level in Norway: adaptation to climate change: perspectives across scales. *Global Environmental Change Part A* **15**, 125–138.

NSW Dept of Environment and Climate Change (2010) NSW coastal planning guideline: adapting to sea level rise. <http://www.planning.nsw.gov.au/LinkClick.aspx?fileticket=VYjmQirQlAk%3d&tabid=177&language=en-US>.

Pini B, Wild River S and McKenzie FMH (2007) Factors inhibiting local government engagement in environmental sustainability: case studies from rural Australia. *Australian Geographer* **38**,161–175.

Preston BL, Smith T, Brooke C, Gorddard R, Measham T, Withycombe G, McInnes K, Abbs D, Beveridge B and Morrison C (2008) 'Mapping climate change vulnerability in the Sydney Coastal Councils Group'. Report prepared for the Sydney Coastal Councils Group and Commonwealth Dept of Climate Change. CSIRO Climate Adaptation Flagship, Canberra.

Preston BL, Brooke C, Measham TG, Smith TF and Gorddard R (2009) Igniting change in local government: lessons learned from a bushfire vulnerability assessment. *Mitigation and Adaptation Strategies for Global Change* **14**, 251–283.

Preston B, Yuen E and Westaway R (2011) Putting vulnerability to climate change on the map: a review of approaches, benefits, and risks. *Sustainability Science* **6**, 177–202.

Renn O (2004) The challenge of integrating deliberation and expertise: participation and discourse in risk management. In *Risk Analysis and Society: An Interdisciplinary*

Characterisation of the Field. (Eds T MacDaniels and M Small) pp. 289–366. Cambridge University Press, Cambridge.

Renn O and Schweizer P-J (2009) Inclusive risk governance: concepts and application to environmental policy making. *Environmental Policy and Governance* **19**, 174–185.

Renn O, Klinke A and van Asselt M (2011) Coping with complexity, uncertainty and ambiguity in risk governance: a synthesis. *AMBIO: Journal of the Human Environment* **40**, 231–246.

Shaw A, Sheppard S, Burch S, Flanders D, Wiek A, Carmichael J, Robinson J and Cohen S (2009) Making local futures tangible: synthesizing, downscaling and visualizing climate change scenarios for participatory capacity building. *Global Environmental Change* **19**, 447–463.

Smith TF, Preston B, Gorddard R, Brooke C, Measham TG, Withycombe G, Beveridge B and Morrison C (2008a) 'Regional Workshops Synthesis Report: Sydney Coastal Councils' vulnerability to climate change. Part 1'. Report prepared for the Sydney Coastal Councils Group. CSIRO Climate Adaptation Flagship, Melbourne.

Smith TF, Preston B, Gorddard R, Brooke C, Measham TG, Withycombe G, Beveridge B and Morrison C (2008b) 'Case studies of adaptive capacity: systems approach to regional climate change adaptation strategies'. Report prepared for the Sydney Coastal Councils Group. CSIRO Climate Adaptation Flagship, Canberra.

Stern N (2007) *The Economics of Climate Change: The Stern Review.* Cambridge University Press, New York.

Tàbara JD, Dai X, Jia G, McEvoy D, Neufeldt H, Serra A, Werners S and West JJ (2010) The climate learning ladder: a pragmatic procedure to support climate adaptation. *Environmental Policy and Governance* **20**, 1–11.

Wild River S (2005) Enhancing the sustainability efforts of local governments. *International Journal of Innovation and Sustainable Development* **1**, 46–64.

Wild River S (2006) Australian local government attempts to deliver beneficial environmental outcomes. *Local Environment: International Journal of Justice and Sustainability* **11**,719–732.

Part III

Governance and risk

12 UNCERTAINTY AND AMBIGUITY IN ENVIRONMENTAL GOVERNANCE: WATER QUALITY IN GREAT BARRIER REEF CATCHMENTS

Bruce Taylor, Tabatha Wallington and Cathy Robinson

Environmental regulation aims to minimise harm to the natural environment or reduce the risk of future harm by structuring patterns of natural resource use and management. Contemporary environmental regulation increasingly relies on an assemblage of practices and instruments. These practices can include prescriptive regulation with sanctions for non-compliance, cooperatively agreed standards or performance measures and the provision of financial incentives and information to encourage voluntary action by land managers. The efficacy of these more cooperative practices often 'depends on affecting the behaviour and the expectations' of a multiplicity of intermediate actors (Majone 1999, p. 5).

In Australia, non-government actors such as regional natural resource management bodies,[1] agri-political organisations and other NGOs are becoming more involved in both the development of these instruments and their implementation through negotiation and collective action (Lane and Morrison 2006; Taylor 2010; Blackstock *et al.* 2007). Because of this, regulation forms part of a broader process of environmental governance, in which there are varying degrees of cooperation (and competition) among the regulators and the traditional subjects of regulation.

Actors in these regulatory contexts must already contend with the limits of scientific knowledge of complex problems, such as managing diffuse water quality from agricultural landscapes. There are also uncertainties about the instrumental efficacy of proposed courses of action. The increasing diversity of policy instruments and involvement of non-government interests present manifestly different questions of institutional, political and operational dimensions of regulation (Rhodes 1997). These concern ambiguities which arise from relational and pragmatic considerations and are distinct from questions of scientific uncertainty (Forester 1989, 1993).

This chapter examines relationships between scientific uncertainty, on the one hand, and political, ideological and moral ambiguity on the other in the development of cooperative standards for regulating water quality in agricultural landscapes. In particular, we examine the implementation of the Reef Water Quality Protection Plan, an overarching policy framework that includes strategies to set water quality condition targets and accelerate the adoption of improved farming practices in the catchments of Australia's Great Barrier Reef. We focus on the Wet Tropics region in far north Queensland. In this region, cattle grazing and intensive sugarcane and horticulture production pose environmental risks from sediments, nutrients and pesticides to coastal and marine ecosystems of the Great Barrier Reef. Farmer interests, governments and other groups have pursued both cooperative and unilateral strategies of regulation in the region for the last decade.

In the next section, we define and explain the concepts of uncertainty and ambiguity before describing the study focus and methods. Our analysis identifies several kinds of uncertainty and ambiguity as characteristic of interaction among government, farming and other interests in policy implementation processes. We then consider the influence of these factors on the effective regulation of farmers' conduct and on the organisations involved in implementation. This chapter builds on insights gained from previous studies that focused on the views of farmers or farmer interests in the implementation of pollution control and agri-environmental schemes (Ward *et al.* 1995; Kaljonen 2006; Blackstock *et al.* 2007; Dupraz *et al.* 2009). This study differs in engaging farming and other actors at the next tier of social organisation, through agri-political organisations (referred to hereafter as farmer industry groups), in environmental governance.

Uncertainty and ambiguity in environmental planning

Uncertainty, in essence, relates to informational deficits, doubts about factual claims or a lack of knowledge about a phenomenon (Forester 1993; Dewulf *et al.* 2005). Common concerns in environmental planning include uncertainty about the cause, extent or magnitude of environmental degradation related to particular management regimes (McDonald and Roberts 2006). Uncertainties are also associated with the efficacy of a proposed course of action to influence behaviour or remedy environmental harm (Pahl-Wostl 2002; Robinson *et al.* 2009). What is less recognised and understood is the effect of ambiguity in situated environmental planning and regulation problems. Ambiguity is normative, generated when expected or acceptable patterns of behaviour are unclear or when meanings associated with issues are contested (Stirling 1998; Wynne 1992). As such, it is 'irreducible and fundamentally different' from technical or cognitive uncertainty, which results from not having enough information on which to make decisions or set thresholds (Forester 1993, p. 89).

Where there are doubts about legitimacy, expressive or meaning-constituting claims of different actors it gives rise not to uncertainty, but to ambiguity (Forester 1993). In natural resource management problems, ambiguity is also defined as the 'simultaneous presence of multiple frames of reference to understand a certain phenomenon' (Dewulf *et al.* 2005, p. 116). These situations require policy-makers and participants to address the 'contingent realities of policy and politics' in addition to questions of instrumental scientific rigour (Pahl-Wostl 2002, p. 399). Forester distinguishes between the needs of uncertainty and those of ambiguity in the following way:

> Wondering about uncertainty, we look for evidence. Wondering about ambiguity we look for precedent; tradition; a source of legitimacy; a consensually based interpretation; an appropriate, fitting response (Forester 1993, p. 90).

Forester (1993) holds that, when planners and policy-makers are confronted with problems of ambiguity and uncertainty, a common response is to confuse the two or to reduce ambiguity to questions of scientific certainty. In this way, competing interpretations of fairness or responsibility may be misdiagnosed or seen as an information deficit that can be remedied through the collection of more data, rather than through practical social action of discussion or bargaining. In the confusion, participants' attention is 'distracted both from the historical and political basis of conflicting and shifting goals and interests, and from the reconstructive moral, legal and political processes that may yet address such issues' (Forester 1993, p. 89). This tendency to reduce ambiguity to uncertainty is heightened by the strong presence of instrumental rationality in the technical and administrative arrangements that characterise environmental policy processes (Healey 1993; Rydin 2003). Indeed, an overemphasis on 'proving' scientific certainty in cases of water planning can lead to asymmetries of understanding, exacerbating the potential for conflict (Steyaert and Jiggins 2007).

Often the ambiguities associated with planning and regulation have a structural basis in the multiple and competing ideologies that underpin contemporary environmental policy processes. Lockie and Higgins (2007) describe agri-environmental policy settings in Australia as hybrid. This hybridity, they argue, is a product of governments pursuing apparently contradictory policy goals by embedding strategies for community participation and empowerment within models of economically rational regulation.

Related sources of ambiguity include the often unacknowledged and competing political motivations behind target setting and policy objectives. Ward *et al.* (1995), writing about the UK *Control of Pollution Act 1974*, described how the 'control objectives' of the regulations were left unclear. In this instance, they pointed to the emerging value-based conflict between farmers and environmentalists. Because of the dominance of agricultural values at that time, the legislated water quality objectives were qualified by considerations of agricultural efficiency and productivity, meaning that farmers were less likely to be sanctioned for polluting waterways. Ambiguity about the objectives provided a political and regulatory buffer for the farming community.

More broadly, the language contained in environmental plans is a potential source of ambiguity since the 'linguistic content of the plan or policy becomes a resource to justify regulatory or investment decisions' (Rydin 2003, p. 84). The language in plans often reflects the compromises or conflicts involved in their creation, giving rise to multiple and contradictory discourses (Rydin 2003; Healey 1993). Where responsibility for implementing regulations sits outside the authority that prepared the plan, there is a greater chance of alternative interpretations and the exercise of regulatory discretion by the subsidiary agency (Rydin 2003).

The introduction of new legislation can be another source of ambiguity in environmental management. Lane *et al.* (2009a) describe the introduction of natural resource management legislation in South Australia as a state-based reform to assist the implementation of a national environmental program. The legislative changes altered the status and roles of various actors, including governments and regional natural resource management boards. In the years during which the new legislation was drafted – and the ensuing implementation period – transaction costs were increased, the potential for conflict was heightened and role redundancy was evident among participants (Lane *et al.* 2009a).

These multiple sources of ambiguity – competing values and political motives, the language of plans and policies, regulatory discretion from devolved responsibility, and legislative change – are features present in the policies and processes associated with the co-regulation arrangements of the Reef Plan. The cooperative and target-based strategies adopted in this plan are outlined in the following section.

Research focus and methods

Targets as co-regulation

We described above how governments, in the practice of contemporary environmental governance, engage the cooperation of other interests to employ a diverse suite of regulatory instruments. Adopting such a portfolio approach has several benefits including the avoidance of an implementation deficit resulting from a purely prescriptive approach (Maddock 2004; Osborn and Datta 2006; Martin *et al.* 2007). It is also seen as contributing towards resolving conflicting policy imperatives (Moss 2004; Carter *et al.* 2005) which are part and parcel of neoliberal policy settings in Australia. One particular regulatory strategy is that of cooperatively developed standards or targets, tailored to local conditions (Ward *et al.* 1995; Sabel *et al.* 2000; Brouwer *et al.* 2003). In our study, these cooperatively developed targets and standards play a central but partial approach to co-regulation of agriculture in the Great Barrier Reef. The policy initiative we focus on is the implementation of the Reef Water Quality Protection Plan (Reef Plan). Co-signed by the Queensland and Commonwealth governments in late 2003, the Reef Plan aimed to 'halt and reverse the decline of water quality entering the Great Barrier Reef lagoon by 2013'. It offered nine strategies to achieve that aim.[2] The two strategies of interest to this research were the development of priorities and resource condition targets, and the promotion of self-management approaches for changes to farming practices to reduce the export of nutrients, sediments and pesticides to the Reef lagoon. These strategies represented an implied willingness by governments to seek change through, at least in the first instance, more cooperative means. On the first strategy, the Plan stated that:

> [Government agencies will] work in partnership with Regional NRM Bodies to determine water quality environmental values and objectives, and to develop aspirational and short-term resource condition and management action targets that reflect the goal of the [Reef Plan] Completed by 1 July 2005 (2003, p. 24).

These targets were, in effect, collective environmental performance standards for land users and resource managers in the catchment. That is, the regional- or catchment-level target did not pertain to individual behaviour, nor was accountability for achieving the target directly attributable to any one actor, public or private. The target itself pertained to a desired, quantified environmental condition for a specified area such as a catchment unit. The development and attainment of that condition was predicated on three non-coercive requirements. The first was that targets were informed principally by scientific knowledge of thresholds of ecosystem harm. Indeed, public science institutions played a significant role in facilitating and advising on these targets at a catchment level and across the Reef (Kroon *et al.* 2009). The second was that the target was consensually endorsed by government, industry and community stakeholders. The final requirement was that information and financial support be provided to farmers, via regional bodies, to encourage change in practices to meet the target. However, encouraging land management practice change was confounded by difficulties in deciding where and how to focus remediation actions, and the efficacy of those actions (Productivity Commission 2003).

On the last requirement, the Reef Plan presented 'self-management approaches' as another important strategy for implementation. The plan's authors recognised that its success would be 'heavily dependent on significant uptake of best management practices in high-risk catchments' and on its promotion through the regional planning and investment process in partnership with industry bodies (Commonwealth of Australia and State of Queensland 2003, p. 12). The following quote illustrates how strategies of self-management were positioned in a broader frame of risk management that included the potential for more formal, prescriptive means:

> Appropriate use of the wide range of existing regulatory powers is essential to complement and support self-management and co-operative partnership approaches. Regulation is an efficient and effective way of affirming the minimum acceptable performance, providing certainty to all stakeholders and ensuring that the efforts and competitiveness of good performers are not

> undermined by those who do not initiate action themselves. Some regulatory actions should be implemented immediately. Others should be implemented at a later date where there is a risk that voluntary approaches will fail to deliver significant water quality improvements (Commonwealth of Australia and State of Queensland 2003, p. 21).

This broadly cooperative approach can be seen as an example of co-regulation of environmental policies in agriculture (Darbas *et al.* 2008), albeit one buttressed with the 'stick' of more formal prescription. Behind these strategies was the intent to deliver a consistent but regionally tailored intervention in diffuse water quality pollution across the six Reef regions, including the Wet Tropics. It was hoped that the approach would avoid problems of regulatory standardisation across diverse biophysical environments, production systems, commodity types and a wide array of interests. These problems of standardisation extend even to the individualised and strongly contextualised character of farming practices on individual farms (Kaljonen 2006).

Research methods

The analysis was constructed through the interpretation of 13 face-to-face, semi-structured interviews conducted by the first author during 2008. Interview participants were government, regional body and farming interests participating in water quality planning and implementation in the Wet Tropics region (including the Tully-Murray River catchment within the Wet Tropics region) and more broadly in Reef Plan policy discussions.

Interview participants included senior policy advisors to state-wide agri-political organisations for sugarcane, horticulture and grazing, and scientific and policy advisors within the Commonwealth government. At more local levels, operational and field staff of agri-political organisations were interviewed, as were regional body employees and catchment planning committee members. Workshop discussions with farming interest groups and regional NRM planners were also conducted by the authors during 2008 and 2009. Those discussions focused on procedural and political issues in the development and implementation of water quality targets. Interpretations relied on the authors' participant observations of planning committees, science forums, public forums and industry reference groups convened by regional bodies between 2006 and 2009.

The following analysis identifies the types of uncertainty and the forms of ambiguity in interactions around water quality regulation under the Reef Plan. We illustrate these using excerpts from the interviews. The chapter concludes by discussing the implications of these findings for agri-environmental regulation and for environmental governance more broadly.

Uncertainties and their effects

The types of uncertainty raised by participants focused on instrumental concerns regarding the establishment of accurate thresholds of ecological harm in marine and coastal environments. The information available from environmental monitoring and modelling exercises in Reef catchments was generally considered by government and industry participants to be 'developing'. This system-level uncertainty was clear in comments by a senior government representative, regarding the timing and nature of environmental impacts from diffuse pollution and the ability to predict those impacts:

> We believe that there will be an impact … of ever-increasing inputs … some time down the track but the Great Barrier Reef is a huge complex system and it's very hard to know when we're going to get to a point where something will start falling over.

The policy environment and the political importance of the Reef Plan meant that scientific uncertainty needed to be weighed against urgency for action and available farming and policy technologies. On the one hand, science was inconclusive as to the sources of declining water quality. On the other hand, short implementation timelines prompted governments and regional groups to try and implement some 'no regrets' actions to meet the Reef Plan's objectives (Robinson *et al.* 2009). Relative unknowns about the capacity of the natural system to buffer prolonged exposure to increased nutrient, pesticide and sediment concentrations also generated uncertainty about how the proposed thresholds might translate into meaningful and measurable targets in plans:

> Reef science is probably the most challenging because often your actions on ground are very divorced from the measures that you're talking about which are highly uncertain at the end of the catchment. Intuitively, people are very suspicious of them because they know they're so uncertain.

This comment, by a regional body project manager, suggests that the perceived lack of credible science informing the target-setting process raised issues of trust in the policy agenda among participants. The effect of this type of uncertainty can be better understood when considered in light of comments made by a senior agri-political representative engaged in policy-level debates on the science behind targets:

> So the farmer has no way of knowing whether what he does today is going to reduce the impact down at the end of the estuary by riparian parts per billion or not and, indeed, if parts per billion down at the estuary increase or decrease the farmer has no way of knowing whether it's because of his action or his neighbour's action or the catchment action ... there is this gap between what I do on my farm and achieving a scientific measurement 15 or 20 km downstream.

Here, the issue about biochemical traceability of pollutants, from individual farms to in-stream or marine environments, extends beyond the instrumental efficacy of practice change on farms or issues of scientific capability. These comments start to raise questions about the ability to determine farmers' performance in terms of accountability or legal responsibility for the detection of spatially or temporally distant pollutants. This is a challenge commonly recognised in regulating diffuse sources of pollution (McDonald and Roberts 2006).

More practically, uncertainty about the efficacy of proposed changes to nutrient management practices on farms – by and large, reductions in the volume of nitrogenous fertilisers applied to crops – remained a persistent issue for farmers and their representatives. For instance, during drafting of the catchment-level water quality improvement plans there were debates about the calculation of profitability and productivity impacts of reducing fertiliser use. Catchment planning committee deliberations on water quality targets for receiving waters at the river mouth were regularly 'interrupted' by pragmatic concerns over how to encourage viable and 'sensible' approaches to establishing the utility of improved farm practices.

During workshop discussions, agricultural stakeholders emphasised feasibility and equity issues surrounding the proposed objectives and the importance of incorporating local industry knowledge (Robinson *et al.* 2009). A significant implementation gap was seen to exist between the conceptual design and actual implementation of modified farming practices. This was illustrated in the comments of an agri-political representative working at the catchment level in the Wet Tropics region:

> ... and just concentrating on application, you know, application methods. That, to some extent, is reasonably easy. It's probably a bit harder to be saying how sure are

> we that the guys [farmers] know how much, how far they can go down, or what alternate products they can use, and what's the science behind it, and why does it work?

Uncertainties about the effectiveness of farm-level action, the scientific resonance of targets and the relationship to a system-wide response were evident. In describing them, connections to legal, political or institutional ambiguities also started to become visible. The next section discusses issues of ambiguity in the co-regulatory environment as it unfolded.

Ambiguities and their effects

Several dimensions of ambiguity were evident from the responses of study participants. They related to the goals and practices of policy implementation, the roles of different players and the rules guiding interaction and political commitment. Perhaps the most pervasive source of ambiguity, from the outset, was the hybridity embodied in the policy design (Lockie and Higgins 2007). A regional body interviewee described how multiple ideologies, including neo-liberal, instrumental and empowerment assumptions in the policy framework, created tensions between competing interpretations of implementation focus:

> Embedded in that I think is a tension between the investors [governments] wanting to buy the most cost-effective outcomes and wanting the regions to be really strategic about what they invest in and where and so forth, versus the regions that are really negotiating a social change process. So for them issues of sort of equity and engagement and so forth are right up there within terms of the immediate biophysical outcomes.

As regional bodies and farming industry groups discussed implementation arrangements, issues of role ambiguity emerged. These focused on whether regional bodies would deliver only voluntary, incentive-based support to farmers or whether they would assist governments to enact statutory nutrient management regulations. Not only roles, but the possible authority status of partners, were questioned:

> [If] it's soft regulation there will be an expectation that regions might have a role in delivering some of that. There's been a bit of discussion between and in the regions and industry fora and they've said 'we're not in the business of [hard] regulation'. But the lines are never quite clear. Regional bodies in other States have some regulatory roles.

Since the Reef Plan required involvement of actors at several scales, it was not uncommon for negotiations involving local industry actors to proceed at different rates from those involving representatives of the same industry groups at state level. This contributed to a situation where, at times, a local agreement on a course of action or principles for implementation could be inconsistent with the direction of debate at head office level. This created ambiguity for farmers in relation to the official positions of their representative organisations and for local industry operatives regarding the legitimacy of any agreements with regional body partners.

Discussions between regional bodies and industry groups during 2007 and 2008 focused on establishing a partnership for accelerating the uptake of improved farming practices – the Reef Alliance. The initial months of deliberation among Alliance partners highlighted ambiguities regarding the rules governing the new cooperative relationship. A policy officer with one of the farming organisations engaged in the deliberations pointed to the multiple sources of ambiguity in the 'rules of the game' governing the more formal relationships between actors:

> … the structures and the processes have been so ambiguous and it's creating difficulties for our partnership because we're getting conflicting messages, because the governance wasn't worked out up front … that creates big issues in terms of how it can move forward … roles and responsibilities, accountability, who's reporting to whom, why they're reporting, what they're reporting on, time frames, all of those things.

Behind these issues of reporting and accountability were more fundamental equity concerns about how funds would be distributed among the parties and who would 'sign the cheques' or set the criteria for farm-level incentive payments. These concerns were directly expressed during agricultural Industry Reference Group meetings, convened by the regional body, as the draft water quality improvement plans were being finalised.[3]

During this period, as implementation arrangements started to consolidate, awareness grew of the potential effects of broader political ambiguities related to governments' commitment to the cooperative process. A senior policy advisor in a Commonwealth government agency remarked:

> We're in an uncertain world with regards to whether the Commonwealth Government is going to continue to support regional arrangements. We've had a new change in people who haven't been party to any of the partnership stuff, complete new swing around as happens often with dealing with Commonwealth departments. There's no certainty that there will be a follow-through.

Exacerbating this situation of national commitment, signs were emerging at the state government level that political action might swing towards new prescriptive regulatory measures. These measures would focus on nutrient management in catchments deemed high-risk. As the potential for this form of regulation had been present in the language of the Reef Plan from the outset, interpretations of the nature, timing and motives behind the most recent signals were varied and destabilising for cooperative relationships. One interpretation from a regional body officer echoed concerns that the state government might usurp cooperatively agreed targets as a performance standard for its prescription, betraying a lack of trust at the political/institutional level:

> People were very nervous about targets and saying well, if we commit to these targets how can we be confident that those targets are not going to be picked up by government and used against us?

A final aspect of ambiguity, consistent with Forester's (1993) and Rydin's (2003) descriptions, was the tendency by the state government and contributing scientific organisations to try to ignore or reduce ambiguity in policy implementation to issues of instrumental and administrative certainty. A farming industry policy advisor argued, with considerable frustration, that the Queensland State Government either actively avoided or seemed incapable of accommodating discussions on implementation. He argued that, instead, the government sought to direct its attention to scientific monitoring of water quality and prescribing mandatory performance standards, and thus sought to reduce complex social and relational issues to a simple technical fix:

> … they're [the State Government] saying well out of all the money we're going to develop some regulations and then we're going to monitor water quality, so again it's this continuation about identifying or proving the problem rather than identifying and proving the solution!

This action by the Queensland government, considered retrogressive and unproductive by farming industry groups, was seen as counter to the progress then being made between industry groups and regional natural resource management bodies through the Alliance. This progress involved detailed and extensive development of schedules of agreed farming practices to improve water quality, voluntary and incentive-based mechanisms to implement them, and monitoring and appraisal frameworks to assess individual farmer and catchment-level progress in implementation. These two sets of actions, operating simultaneously, did little to stabilise the confidence of farming and regional actors in the future legitimacy of the emerging partnership arrangements.

Implications for theory and practice

There are several themes from the analysis that warrant more exploration. First, we discuss the value of uncertainty and ambiguity as analytical concepts then offer an insight into ambiguity as both a symptom and an influential force in strategies for cooperative environmental regulation and governance. Finally, we suggest that insights from research of this kind can inform more practical tasks of policy instrument selection and appraisal.

The first theme of the analysis relates to uncertainty and ambiguity as analytical concepts. Our findings illustrate Dewulf *et al.*'s (2005) argument that a conceptual separation of uncertainty and ambiguity is easier than an analytical one. Uncertainties may be compounded by ambiguities that influence actors' perceptions of a given situation or proposal. The ties between descriptive issues of scientific and instrumental uncertainty in target-setting, and normative issues of trust, accountability and political risk, are clearly evident in the planning and policy debates described above. Uncertainties around the efficacy and impact of farm-level practice change are exacerbated by ambiguities concerning the commitment of public resources to share the costs of implementation, and the future policy environment, which farmers and their representatives tend to assume might be hostile to their goals. To manage environmental risks effectively, both these normative and evaluative/cognitive dimensions need to be considered (Klinke and Renn 2002).

The second theme is that attempts at co-regulation can be both a source of and a response to ambiguity in environmental governance. Instrumental uncertainties are understood to be characteristic of regulation in complex science-policy arenas (Pahl-Wostl 2007; Eberhard *et al.* 2009). However, ambiguity appears to be generated during attempts by governments and other actors to move towards new, more cooperative approaches. In part, this reflects the 'drama of mediation', where deep-seated values and value differences are played out and elements of power plays are exposed (Forester 1999). The new levels of cooperation require clarification of multiple interpretations and meanings, dynamic role changes, the rules that govern their interaction and the social and political implications of these changes (Forester 1993). State and Commonwealth governments contributed significantly to the ambiguity experienced by regional and local actors in developing water quality targets. This, in part, stemmed from governments devolving several critical implementation responsibilities to regional, industry and local actors who were not signatories to the Reef Plan. This allowed regional bodies and industry groups to exercise considerable levels of discretion. Ambiguity also stemmed from governments seeking to spread the risk of policy implementation by adopting multiple regulatory strategies consistent with hybrid regimes (Lockie and Higgins 2007). One such strategy involved government persisting in its oversight role by setting the problem framework and the rules under which different interests engaged (Bell and Park 2006). Regulation in this setting was characterised by ambiguity, conflicting signals in the language of plans and policies, and poorly defined or contested rules. According to participants, the resultant blurring of responsibility and accountability placed considerable strain on the development and implementation of co-regulatory arrangements (Stoker 1998).

The Queensland government's signal in 2008 of its intention to shift back to more formal mandatory regulatory controls[4] can be seen as an attempt to reduce the undesirable ambiguity of alternative interpretations and local and regional discretion in implementation. These signals were evident in the Reef Plan statement that mandatory regulations provided more 'certainty' for all participants (Commonwealth of Australia and State of Queensland 2003, p. 21). The signal to regulate was accompanied by the release of a Scientific Consensus Statement by the Queensland government (State Government of Queensland 2008), consistent with Forester's (1993) view that planners and bureaucrats tend to reduce ambiguity to questions of scientific or instrumental uncertainty. The focus on quantification of targets at catchment level distracted discussions from the 'reconstructive moral, legal and political processes' (Forester 1993, p. 89) that may have addressed concerns articulated by farmer industry group representatives; concerns such as the inefficient use of public resources and the unfair burdens placed on farmers. This argument has also been expressed in wider-reaching reviews of environmental regulatory practice in Australia (Martin *et al.* 2007).

Last, there is the question of how our analysis might inform practical regulatory design and appraisal. Gunningham and Sinclair's (2005) work on policy instruments in diffuse water quality pollution (and other) regulatory contexts proposed criteria for instrument selection. They suggested policy-makers might usefully adopt criteria of cost-effectiveness, equity and political acceptability in their deliberations. As we have demonstrated, an understanding of ambiguity helps to inform a better understanding of the distributional, political and relational questions of environmental co-regulation. The type of analysis presented in this chapter could thus make a significant contribution to a richer and more realistic application of political acceptability and equity criteria.

We have argued that a greater appreciation of normative ambiguity and more instrumental understandings of uncertainty in cooperative governance, together with a richer appreciation of their interdependencies, can set the stage for actors to identify and craft politically and socially acceptable compliance outcomes. Our analysis suggests that such outcomes are likely to provide greater levels of trust and cooperation between governments and other actors. While not sufficient in themselves to mitigate environmental risk from farming activities, this more pragmatic approach creates conditions where new instrumental knowledge or alternative regulatory practices can be more meaningfully debated, deployed and revised through the processes of collective action already underway in regions and catchments.

Endnotes

1 These regional bodies are, principally, government-funded multi-stakeholder organisations with planning and coordination responsibilities for natural resource management. There are regional bodies established in each of the 56 natural resource management regions in Australia. In some state and territory jurisdictions the bodies have statutory and regulatory powers, in others they do not (Lane *et al.* 2009b; Robins and Dovers 2007).

2 The nine broad implementation strategies presented in the Reef Plan include self-management approaches, education and extension, economic incentives, planning for

natural resources and land use, regulatory frameworks, research and information sharing, partnerships, priorities and targets, and monitoring and evaluation.

3 Participant observations, Industry Reference Group meetings, Tully (December 2007, April 2008) and Innisfail (October 2009).

4 This was realised through the introduction of the Great Barrier Reef Protection Amendment Bill 2009 which established requirements for farmers in high-risk catchments (for nutrient export) to prepare and submit farm-level Environmental Risk Management Plans to the Queensland State Government for assessment.

Acknowledgements

We gratefully acknowledge the participation of and insightful discussions with regional body planners and CEOs, representatives of farming industry groups (Growcom, Agforce and Canegrowers) and government representatives, particularly those involved in the Wet Tropics regional NRM process and Tully-Murray Water Quality Improvement Plan. The research was funded by the Australian government's Marine and Tropical Sciences Research Facility and CSIRO Water for a Healthy Country Flagship.

References

Bell S and Park A (2006) The problematic metagovernance of networks: water reform in New South Wales. *Journal of Public Policy* **26**, 63–83.

Blackstock K, Brown K, Davies B and Shannon P (2007) Individualism, cooperation and conservation in the Scottish farming communities. In *Rural Governance: International Perspectives*. (Eds L Cheshire, V Higgins and G Lawrence) pp. 191–208. Routledge, London.

Brouwer F, Heinz I and Zabel T (Eds) (2003) *Governance of Water-related Conflicts in Agriculture: New Directions in Agri-environmental and Water Policies in the EU*. Kluwer Academic, Dordrecht.

Carter N, Kreutzwiser RD and de Loe RC (2005) Closing the circle: linking land use planning and water management at the local level. *Land Use Policy* **22**, 115–127.

Commonwealth of Australia and State of Queensland (2003) 'Reef Water Quality Protection Plan'. Qld Dept of Premier and Cabinet/Dept of Environment, Water, Heritage and the Arts, Australian Government, Brisbane. <http://www.reefplan.qld.gov.au/library/pdf/reefplan.pdf>.

Darbas T, Reeve I, Farquharson R and Graham S (2008) Co-regulation and cotton: governance of natural resource management in the Australian cotton industry. *Australasian Journal of Natural Resources Law and Policy* **12**, 87–114.

Dewulf A, Craps M, Bouwen R, Taillieu T and Pahl-Wostl C(2005) Integrated management of natural resources: dealing with ambiguous issues, multiple actors and diverging frames. *Water Science and Technology* **52**, 115–124.

Dupraz P, Latouche K and Turpin N (2009) Threshold effect and co-ordination of agri-environmental efforts. *Journal of Environmental Planning and Management* **52**, 613–630.

Eberhard R, Robinson CJ, Waterhouse J, Parslow J, Hart B, Grayson R and Taylor B (2009) Adaptive management for water quality planning: from theory to practice. *Marine and Freshwater Research* **60**, 1189–1195.

Forester J (1989) *Planning in the Face of Power*. University of California Press, Berkeley.

Forester J (1993) *Critical Theory, Public Policy and Planning Practice: Toward a Critical Pragmatism*. State University of New York Press, Albany.

Forester J (1999) Dealing with deep value differences. In *The Consensus-building Handbook.* (Eds L Susskind, S McKearnan and J Thomas-Larmar) pp. 463–493. Sage, London.

Gunningham N and Sinclair D (2005) Policy instrument choice and diffuse source pollution. *Journal of Environmental Law* **17**, 51–81.

Healey P (1993) The communicative work of development plans. *Environment and Planning B* **20**, 83–104.

Kaljonen M (2006) Co-construction of agency and environmental management: the case of agri-environmental policy implementation at Finnish farms. *Journal of Rural Studies* **22**, 205–216.

Klinke A and Renn O (2002) A new approach to risk evaluation and management: risk-based, precaution-based and discourse-based strategies. *Risk Analysis* **22**, 1071–1094.

Kroon FJ, Robinson CJ and Dale AP (2009) Integrating knowledge to inform water quality planning in the Tully-Murray basin, Australia. *Marine and Freshwater Research* **60**, 1183–1188.

Lane MB and Morrison TH (2006) Public interest or private agenda? *Journal of Rural Studies* **22**, 232–242.

Lane MB, Haygreen A, Morrison TH and Woodlands J (2009a) Will regionalisation achieve integrated natural resource management? Insights from recent South Australian experience. In *Contested Country: Local and Regional Natural Resources Management in Australia.* (Eds MB Lane, CJ Robinson and B Taylor) pp. 59–74. CSIRO Publishing, Melbourne.

Lane MB, Taylor B and Robinson C (2009b) Introduction: contested country – regional natural resource management in Australia. In *Contested Country: Local and Regional Natural Resources Management in Australia* (Eds MB Lane, CJ Robinson and B Taylor) pp. 1–12. CSIRO Publishing, Melbourne.

Lockie S and Higgins V (2007) Roll-out neoliberalism and hybrid practices of regulation in Australian agri-environmental governance. *Journal of Rural Studies* **23**, 1–11.

Maddock TA (2004) Fragmenting regimes: how water quality regulation is changing political-economic landscapes. *Geoforum* **35**, 217–230.

Majone G (1999) The regulatory state and its legitimacy problems. *West European Politics* **22**, 1–24.

Martin P, Bartel R, Sinden J, Gunningham N and Hannam I (2007) 'Developing a good regulatory practice model for environmental regulations impacting on farmers'. Overview research report prepared for the Australian Farm Institute and Land and Water Australia, Sydney.

McDonald G and Roberts B (2006) SMART water quality targets for Great Barrier Reef catchments. *Australasian Journal of Environmental Management* **13**, 95–107.

Moss T (2004) The governance of land use in river basins: prospects for overcoming problems of institutional interplay with the EU Water Framework Directive. *Land Use Policy* **21**, 85–94.

Osborn D and Datta A (2006) Institutional and policy cocktails for protecting coastal and marine environments from land-based sources of pollution. *Ocean and Coastal Management* **49**, 576–596.

Pahl-Wostl C (2002) Towards sustainability in the water sector: the importance of human actors and processes of social learning. *Aquatic Sciences* **64**, 394–411.

Pahl-Wostl C (2007) The implications of complexity for integrated resources management. *Environmental Modelling and Software* **22**, 561–569.
Productivity Commission (2003) 'Industries, land use and water quality in the Great Barrier Reef catchment'. Research report. Productivity Commission, Canberra.
Rhodes R (1997) *Understanding Governance*. Open University Press, Buckingham.
Robins L and Dovers S (2007) Community-based NRM boards of management: are they up to the task? *Australasian Journal of Environmental Management* **14**, 111–122.
Robinson CJ, Taylor BM, Pearson L, O'Donohue M and Harman B (2009) A SMART assessment of water quality partnerships in Great Barrier Reef catchments. *Australasian Journal of Environmental Management* **16**, 84–93.
Rydin Y (2003) *Conflict, Consensus and Rationality in Environmental Planning: An Institutional Discourse Approach*. Oxford University Press, Oxford.
Sabel C, Fung A and Karkkainen B (2000). *Beyond Backyard Environmentalism*. Beacon Press.
State Government of Queensland (2008) 'Scientific consensus statement on water quality in the Great Barrier Reef'. Reef Water Quality Protection Plan Secretariat, Brisbane.
Steyaert P and Jiggins J (2007) Governance of complex environmental situations through social learning: a synthesis of SLIM's lessons for research, policy and practice. *Environmental Science and Policy* **10**, 575–586.
Stirling A (1998) Risk at a turning point? *Journal of Risk Research* **1**, 97–110.
Stoker G (1998) Governance as theory: five propositions. *International Social Science Journal* **50**, 17–28.
Taylor BM (2010) Between argument and coercion: social coordination in rural environmental governance. *Journal of Rural Studies* **26**, 383–393.
Ward N, Lowe P, Seymour S and Clark J (1995) Rural restructuring and the regulation of farm pollution. *Environment and Planning A* **27**, 1193–1211.
Wynne B (1992) Uncertainty and environmental learning: reconceiving science and policy in the preventive paradigm. *Global Environmental Change* **2**, 111–127.

13 CHOICE EDITING FOR THE ENVIRONMENT: MANAGING CORPORATE RISKS

Jane Dixon and Cathy Banwell

> This report is about the good life: the good life that is possible if people, business and government all share responsibility to open the way to new solutions (Sustainable Consumption Roundtable 2006, p. i).

Integral to shared responsibility for perpetuating and sustaining 'the good life' are corporations, and in this chapter we examine a particular sustainable development approach by a major international supermarket chain, Marks and Spencer Group (Marks and Spencer). We focus on this corporation because it has long been viewed as a role model for other supermarket chains, as exemplified by its early championship of corporate social responsibility (CSR). Among other activities, CSR has been used by retailers for over 20 years to reconfigure production and logistics processes so as to meet the constantly evolving expectations of stakeholders and investors. It has also added a moral authority missing from the routine economic exchange at the heart of retail capital accumulation. CSR is generally viewed as helping retailers to 'be better placed to provide long-term growth and financial security for [all] stakeholders and to maintain and enhance their market position' (Jones *et al.* 2005, p. 891).

CSR applies to an ever-expanding number of issues ranging from improving animal welfare and public health to fair labour practices. One of the earliest issues for attention was environmental sustainability. Marks and Spencer is among the top 10 UK retailers undertaking CSR, and concern for the environment is one of its five key pillars of sustainability. In common with other large firms, Marks and Spencer is attempting to reduce its ecological footprint by insisting that its producers change their environmental resource use practices and by reducing resource use in its stores and throughout its logistics systems. Marks and Spencer is also attempting something new – being seen to support 'green consumerism' beyond the obvious stocking of 'green' labelled products.

Within a context of dire predictions regarding planetary overload from human activity, the spotlight is focused on encouraging affluent consumers to use their consumption practices in socially and environmentally responsible ways. A cultural expectation is being fostered by social movements, ecologists, religious bodies and ethical investment firms for consumers to consciously consider the environment and future generations of people and animals when they use resources. Participants in a consumer forum sponsored by the UK's Sustainable Consumption Roundtable (SCR) cited a range of factors that they considered barriers to changing their resource use patterns. They claimed that:

- we are creatures of habit, reluctant to make changes that challenge our routines;
- we often lack access to facilities like doorstop recycling or good public transport;
- we perceive sustainable options to be expensive and niche (SCR 2006, p. 9).

One part of the multi-faceted research program supporting the Roundtable's work involved an examination of whether consumers will act unaided on their sustainability concerns. Analysis of the driving forces behind 19 product innovations showed that people were more likely to consume in a sustainable manner if led by business and/or government. It was concluded that:

> ... choice editing for quality and sustainability by regulators, retailers and manufacturers has been the critical driver in the majority of cases. Providing information failed to get more than a minority of people buying [for example] the most energy-efficient dishwashers (SCR 2006, p. 21).

The report supported more theoretical analyses which suggest that both everyday and status-oriented forms of consumption are shaped by a variety of 'policies, campaigns and practices that enlist ordinary people into broader projects of social change' (Barnett *et al.* 2005, p. 23). They do not arise spontaneously but are enacted through the alignment of people's consumption practices with the socio-cultural, economic and physical imperatives of the day (Shove 2003; Shove *et al.* 2009; Gibson *et al.* 2011). This establishes the context for our consideration of corporate 'choice editing'.

Emanating from the UK, choice editing is being promoted as a constituent component of retailer-backed sustainability systems. It has been described as:

> ... shifting the field of choice for mainstream consumers: cutting out unnecessarily damaging products and getting real sustainable choices on the shelves. In the context of high consumer concern, but low levels of action, the idea of integrating the most compelling issues of sustainable development through choice editing makes sense. Consumers benefit from the assurance that the issues they care about are considered, rather than facing the demand that they grapple with those complexities themselves (SCR 2006, p. 16).

In practical terms, retailers narrow product ranges by adopting particular quality standards regarding what is environmentally sustainable, healthy, animal-friendly, socially just and so forth. Through choice editing, institutional actors are attempting to make the sustainable product choice the easy and only choice. In the ideal world, there would be no need for moral decisions because all citizens could assume that their shopping decisions had minimal impact on the biosphere. The concept has the support of public health ecologists who are concerned about the population and environmental health problems which arise through unbridled choice. Lang *et al.* (2009) argued that a different approach to choice is needed in the pursuit of ecological public health and put forward a model to illustrate the spectrum of responses to choice by which consumers are governed. Figure 13.1 illustrates the gulf that they believe exists between current policy support for market-based commodity and services choice, and the directions in which societies need to move to deliver a healthy and sustainable food system. It also proposes that, through their capacity to choose, higher socio-economic groups have the highest ecological impact.

Figure 13.1 shows that the rich have further to move in the direction of sustainable behaviour, but they may also be more amenable to reducing their ecological footprint through buying

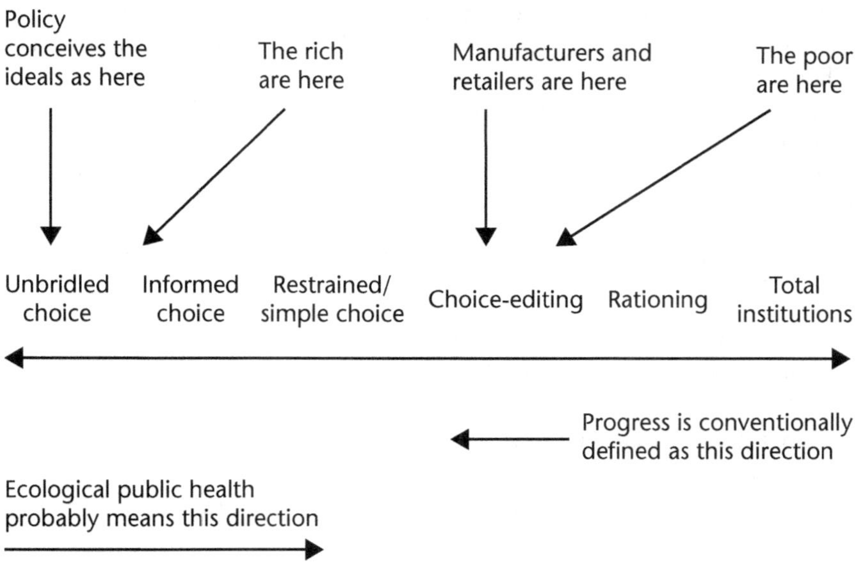

Figure 13.1: Current policy vs the need to move towards more sustainable behaviour.

higher-priced environmentally friendly goods from trusted supermarket chains. In our opinion, the best developed and most impressive example of choice editing with the middle-class consumer in mind is Marks and Spencer's Plan A, a hugely ambitious program to distinguish the chain from its competitors in the UK, European and US markets by becoming 'the world's most sustainable major retailer by 2015' (Marks and Spencer Group 2010).

Corporate response to sustainable food systems: Marks and Spencer Plan A

Plan A was launched in 2007, with Marks and Spencer announcing that it was 'committing to change 100 things over five years – because we've got only one world and time is running out' (Marks and Spencer Group 2010, p. 3). The plan includes three aims: to save the company money, to create brand loyalty among the firm's 21 million customers and to save the planet through acting on five pillars of sustainability – climate change, waste, natural resources, fair partnerships, and health and well-being. These areas of concern about sustainability reflect key concerns of the ecological public health community (Lang *et al.* 2009).

Plan A extends across all the company's production and retailing activities. Marks and Spencer aims to tackle all issues along the supply chain and to be a business that is carbon-positive, wastes nothing, restores the natural environment and improves lives. To this end, the corporation has committed to applying at least one of the five sustainability pillars to its 2.7 billion items by 2020, which will mean working with thousands of companies and tens of thousands of farms (Marks and Spencer Group 2010, p. 5). It will help to create 200 Plan A factories and will enrol 10 000 farmers in the company's Sustainable Agriculture Program. For example, the Marks and Spencer website describes eco-tile factories the corporation has built and is supporting in Sri Lanka.

In terms of addressing sustainable food systems, Marks and Spencer has the following initiatives:

- palm oil standard and use – the company has developed a palm oil audit system and insists that only palm oil produced in a sustainable and responsible way will be used in its house-brand products;
- it will produce and promote functional foods, coupled with innovative convenient product solutions;
- it will offer energetic support for alternative foods, whether organically grown, locally specific niche items or artisanal speciality foods.

Assessing Plan A after three years of operation, the company announced success in the following terms. The strategy had:

- saved the firm £50 million in 2009–10;
- motivated 1 million customers to donate money to charity;
- saved 4 million items of clothing from going to landfill;
- improved store efficiency by 10%;
- reduced food packaging by 16%;
- improved fuel efficiency in transport and logistics by 20% (Marks and Spencer Group 2010, pp. 4–5).

In 2010, the company issued a revised strategy that took Plan A through to 2015 and added 80 new areas of focus. The rationale for the additions was that, 'Customers are aware of their personal impact on the world and ready to take action but most expect us to take a lead and make it relevant to their everyday lives' (p. 5). It claimed consumers still wanted to consume despite the global financial crisis of 2008, but they wanted business and government to do the 'heavy lifting ... by making it easy for them to get involved and to make a difference' (p. 3). The pledge to incorporate at least one of the five pillars of sustainability into all Marks and Spencer products was included in the revised strategy, as was the aim to tailor a personal Plan A for 3 million of the company's 21 million customers by 2015. The company announced that it wanted:

> ... to involve all of our employees in Plan A and help them share the word with family, friends and Marks and Spencer customers, so that's why we're going to offer all eligible employees free home insulation and a home energy monitor, and give all employees a day leave every year to do volunteer work in the communities where they and customers live (p. 7).

For 15% of Marks and Spencer customers and for all staff, acting sustainably and participating in Plan A will become synonymous.

What risks is Plan A attempting to address?

It appears that Marks and Spencer is using Plan A to address two interrelated types of risk. The first concerns medium-term limits to economic growth in a context of global environmental change and resultant questioning of the current economic growth approach to global and national development. The New Economics Foundation (2009), for example, argued that global growth in greenhouse gases increased by an average of 3.5% between 2000 and 2007 (years of unprecedented economic activity in Asia and South America) but dropped during the ensuing global financial crisis. The foundation considered this proved that constant economic growth based on current energy and resource use technology is unsustainable. Publicly available material indicates that Marks and Spencer shows keen awareness that economic growth – whether by nations or companies – fuels greenhouse gas emissions. Like any major corporation, Marks and Spencer faces risks to the legitimacy of its approach to generating profit. By emphasising products that incorporate one of the five pillars of sustainability, the chain is attempting to decouple its money-making activities from global environmental change and to position itself as a socially and environmentally responsible actor.

As a major retailer, Marks and Spencer has traditionally been focused on another form of risk, namely, minimising short-term financial losses as it competes for market share and increased profits every quarter. To increase profits, retailers aim to increase the number of customers who choose to shop at their store and to encourage those customers to buy increasing volumes and types of products (Mitchell and Harris 2005). The firm constantly risks losing custom to competitors, resulting in financial outlays overwhelming profits, leading to overall losses. Reports of reduced profits increase the flow-on risks of loss on the stock exchange.

The heart of the traditional retail model involves attaching significant cultural value to product line and brand choice: companies compete on the size of their product portfolios, their floor space, their changing specials and being the first to introduce novel products. For decades, supermarket chains have made a virtue of the ever-increasing size of their product ranges. To accommodate multiplying food product lines plus new aisles devoted to furniture, clothing and technology, supermarkets established the hypermarket concept. This business model is based on offering a plethora of choices with a range of marketing strategies to encourage hyperconsumption: the here-and-now consumption of large volumes of novel products.

Operating a hyperproduction–consumption business strategy has created two immediate governance risks up and down the supply chain. First, it has meant establishing an ever-expanding number and type of supply chains. But managing the logistics of thousands of supply chains is expensive, particularly as retailers become ever more responsible for the safety, fairness, health and environmental credentials of each product they sell (Anon. 2010). Second, the constant supply of new products involves managing a cascade of consumer risk perceptions at a time when the expansion of hypermarkets has stalled in mature market economies due to growing consumer antipathy for unbridled choice.

Traditionally, consumers have accepted numerous risks when making purchases: physical risks associated with safety, financial risks related to costs, psychosocial risks of feeling unrewarded by the shopping experience, the opportunity cost risks imposed by slow and cumbersome retail formats (Mitchell and Kiral 1999). Most marketing strategies have aimed to reduce these risks by introducing new product lines, designed on the basis of retailers' detailed knowledge of consumer aspirations and problem perceptions. However, countering problems with new products gradually has become less successful as consumers signal an overwhelming desire for convenience, physically and emotionally (Humphery 1998; Dixon 2002). Consumers are now resisting what they see as excessive choice out of concern for environmental sustainability (Humphery 2010; SCR 2006). This threat to corporate profitability actually offers opportunities to corporations which can use their new-found legitimacy from decoupling of profit and environmental change with the in-store recoupling strategy of choice editing.

Choice editing has several merits. It reinforces the strategy of corporate social responsibility, provides a rationale for the more efficient management of supply chains and helps address consumption fatigue brought on by endless decisions over products. It becomes both a corporate promotional tool and a logistics response to managing the risk of losing consumers who desire a simpler approach to shopping and a more ethically fulfilling approach to consuming.

The appeal of Marks and Spencer recoupling sustainability through consumption was indicated by a survey of University of Plymouth students. While not representative of the general population, we can assume that the students were relatively typical of future middle-class consumers. Over 90% of the students surveyed were positive about sustainability. In response to questions about what individuals could do to live a sustainable life-style most nominated behaviour associated with consumer change such as choosing products that were fair-trade, locally produced and/or healthy, and recycling. Some intended to purchase products from companies which are environmentally or socially ethical. Furthermore, while most agreed with radical statements regarding social changes to support sustainability they proffered only minor personal changes focused on their behaviour as consumers (Kagawa 2007). If such views are widespread among university students in the UK, it becomes clear that the Marks and Spencer approach is tapping into consumers' aspirations. Just as some stores are cheap or well managed (Mitchell and Harris 2005), Marks and Spencer is sustainable – it can offer consumers a comfortable way of mitigating their future risk of not being able to live the good life. It is thus plausible that Plan A is integral to the Marks and Spencer strategy to recover its financial and societal reputation, that was undermined when it pulled back from global expansion during the 1990s (Burt *et al.* 2002).

Is sustainable consumption possible in a consumptogenic environment?

Plan A contributes to the profit margins of Marks and Spencer and meets some consumers' need to act for the environment. What are its other potential benefits? We are not going to address the potential of the production-side activities. Much has been written and remains unresolved about commercially driven supply chain responses to environmental degradation. We are more interested in Plan A's attempt to foster sustainable consumption, and whether this can take place at a scale sufficient to affect the environment positively. Conceiving an ethos of sustainable consumption is an interesting exercise given that ideas of restraint, going without, voluntary simplicity or doing more with less, run counter to half a century of celebrating excess, abundance, having it all and living the good life by doing more with more. The latter set of qualities characterises what we term a 'consumptogenic era' – an era when consumption moved from simply fulfilling the basic needs of shelter, food, clothing and mobility and acquired nationalistic, social and moral overtones.

We see three limitations to Marks and Spencer's support for green consumerism, given that it is being fostered in a consumptogenic era. First, Plan A does not caution against unnecessary consumption of the Earth's resources. It is not aimed at limiting the volumes of products traded, or surplus goods at the consumer end. While the strategy of individualised plans approximates informed choice on the spectrum described in Figure 13.1, the company still encourages a sense of unbridled choice when it boasts that it sells 2.7 billion items, all of which will incorporate at least one sustainability pillar. In practice, Plan A becomes a brand-switching platform: move away from unsustainably produced goods and towards Plan A sustainably produced alternatives. It does not challenge the retail business model that profitability lies in competing for brand-loyal, high-volume customers. It does not challenge the individual 'customer is king' orthodoxy.

Our second concern focuses on the implications for the exercise of citizenship. Plan A does not question the way increasing levels of consumerism have become valorised. The conflation of citizenship and consumerism reached new heights during the global financial crisis of 2008–09, when OECD governments used stimulus packages including tax cuts and one-off payments so that citizens could spend more money to resurrect national economic progress. 'Hyperconsumption' is an apt description for the emotional intensity that accompanied consumption as a patriotic duty.

Citizen-oriented hyperconsumption can comfortably sit alongside sustainable consumption when all these practices are ill-defined. In common with most institutions which invoke the word 'sustainable' or 'sustainability', Marks and Spencer does not qualify or enumerate what it means by the term. Rather, it deploys the word as 'a higher-order social goal akin to other such goals like democracy, justice or the rule of law', without acknowledging that it is also 'a complex, contestable challenge spanning generations' (Dovers *et al.* 2008, p. 246). By making 'sustainable choices, easy choices' supermarkets are renegotiating how consumers can imagine and practise their relationships with governments and one another. Complexity is erased.

A Plan A customer can continue to consume at unprecedented levels and believe that if they buy goods labelled 'sustainably produced' they are acting not only to save the economy but to improve the environment. In stocking food products certified organic or ethically or sustainably produced, supermarkets are offering consumers a role in resolving their guilt about the deepening ecological crisis in the food system. Retailer-labelled choices purport to remove the uncertainty or risk about the ecological sustainability of a product and therefore the risk about the future in general. Very few sustainability analyses contain estimates of cumulative impacts, irreversibility (tipping points), complexity and connectivity between problems, or risk and uncertainty (Dovers *et al.* 2008). There is little environmental impact assessment data on the implications of supermarket dominance of global supply chains although this dynamic may be of utmost importance to food system resilience (Erickson *et al.* 2010). Influenced by Campbell's work (2009), we note that corporate leadership to protect the environment and promote sustainability has a troubling limitation: the environmental, health and community feedbacks which arise from ecoconsumption are largely unknown to consumers.

The environmental impact of consumer demands for alternative goods produced in distant locales or fragile ecosystems and supplied through supermarkets is not yet clear. Supermarkets are increasingly involved in the organics sector, but Johnston *et al.* (2009) concluded that corporate organics are riddled with contradictions, based on the fact that commodity fetishisation is maintained. In a context of corporate supply chains with no consumer involvement, externally verified labels and certifications attached to alternative foods 'become a cheap substitute for intimate knowledge of the commodity and producers' (Freidberg 2003, p. 229). In contrast, backyard production maximises this intimate knowledge. It is also present in what may be more environmentally sustainable (due to resource use efficiencies) organic cooperative ventures in which producers and consumers share risks. In corporate supply chains, risk is privatised on behalf of producers and consumers. Financial as well as environmental sustainability risk-sharing is exemplified in the box-scheme approach, when consumers pay up-front (often based on a price that internalises the environmental costs) in anticipation that the farmer will be able to deliver produce of highly accountable provenance (Seyfang 2006). We are not arguing that such schemes can address environmental sustainability on a grand scale, but in such scenarios the 'customer is king' philosophy gives way to a producer–consumer partnership with fewer contradictions. Unbridled choice is not an option, and although global sourcing of sustainable choices may be present in box schemes the very few supplier participants are relatively easy to know and feedbacks are easier to detect.

With the label as the major empowering device the retailer has most of the influence, risking food system resilience. This is our third area of concern. Marks and Spencer has invested an enormous amount of capital in the Plan A portfolio and modus operandi. Staking its corporate reputation on the Plan A brand, the firm micromanages all along the supply chains to ensure nothing goes wrong with the product (e.g. a food safety problem). Its demands, in terms of volume and quality of supply, ensures that only larger suppliers can comply, and the constant push for greater efficiencies can mean that 'communities of producers are pushed to the edge of economic survival' (Sustainability Institute 2003, p. 2). The contraction in suppliers and non-supermarket retailers, largely due to supermarket dominance of supply chains and local food retailing environments, undermines a future food system comprising diverse economic models and producer–consumer engagements which, alongside the protection of biodiversity, is considered by some to be the hallmark of a sustainable food system (Larsen *et al.* 2008).

Diverse producer–consumer subsystems have become truncated as more people enter the workforce from a young age. In describing consumer choice as a mode of governance, Kjaernes *et al.* (2009, p. 3) invoked Giddens' argument that modern societies are characterised by the fact that 'people have no choice but to choose'. The point to add here is that most citizens have no choice but to choose from among the industrialised diversity of the marketplace rather than to choose between home and communally produced goods and market-based goods. This is partly due to a sense that self-reliance or do-it-yourself is not realistic when there are many other pressures – people take account of the fact that they would forgo too much income if they made their own clothes, cooked meals from scratch or constructed leisure opportunities without technological aids. In this context, choice editing provides 'a menu of new, guilt-free shopping options for affluent consumers' (Johnston *et al.* 2009, p. 526).

One way of framing the supermarket role is as an 'eco-moderniser'. Eco-modernisation promises 'environmental protection and resource conservation through means of eco-efficiency and stakeholder engagement' (Milne *et al.* 2009, p. 1241). In reporting on New Zealand business incorporation of sustainability discourse, Milne *et al.* noted a tendency 'to adopt a narrow economic language, standard of judgment, and world view in approaching and utilizing the Earth' (2009, p. 1241). They contrasted this business approach, which privileges the efficiency of resource use, with an ecosystems approach where the focus is on the sustainability gap that opens up whenever resources are used. Marks and Spencer is clearly an eco-moderniser as it uses sustainability discourse to drive an environmental agenda while masking ecological limits.

Despite the contradictions, it is clear that consumers will expect supermarkets to remain major players in the realm of environmental governance and risk management. Their financial and symbolic power in modern food systems ensures that they play a key role in influencing the roles of governments and consumer-citizens. In a context of animal welfare, Kjaernes *et al.* (2008) typified market–consumer–state relations according to three models: the welfare state, *terroir* and quality. They plotted different European countries according to the models. The Scandinavian countries came closest to the first model, because government regulations fostered protected and mass generic markets, and a sense of trust between producers and consumers was brokered through strong regulatory oversight. With their distrust of institutions and the high value placed on familiarity and social control based on small networks, Italy and France aligned with the *terroir* approach. Due to the strong presence of supermarkets, the UK and Netherlands emphasised market solutions and consumer choice; the government's role was to support differentiated or multiple forms of quality. Consumer choice was strongest in the quality model and weakest in the welfare state approach. It is possible that none of these three approaches to food system governance may enhance food system resilience, although the *terroir* approach with its emphasis on bioregional ecosystems, social diversity and consumer choice limited by seasonality, place and tradition rather than novelty, may be revealed as offering the physical environment its greatest chance of protection.

Conclusion

Lang *et al.* (2009) argued that sustainable food systems can emerge only from the symbiotic efforts of commerce, governments and civil society. Marks and Spencer promotes Plan A as a strategy to address the limits to ecosystem services, but we propose in this chapter that it is a strategy fraught with contradictions. At its core is a form of governance called choice editing, the corporate endorsement of relatively environmentally safe products and the rigorous policing of supply chains. In practice, Plan A encourages brand swapping for the environment but does not address the inexorable promotion of market place consumption. It fosters a consumption ethos where the retailer brand and labelling offer peace of mind rather than empowering consumers through full disclosure of environmental costs and benefits, which in turn leaves even more power with the retailer to manage risks. Such dominance of supply chains puts food system resilience at risk. We argue that Plan A is not a corporate-citizen co-sustainability strategy but a corporate brand promotion strategy to minimise financial risks in the near term and to secure a future for corporate wealth creation in the mid term if environmental change escalates.

Without conditions being specified, environmental sustainability strategies can be vague and unaccountable whether practised by governments, firms or households. A lack of clarity within the general population allows the concept of sustainability to be used to suit a range of purposes; we suggest that fluid discursive practices arguably pose the biggest risk to environmental sustainability. Ironically, if households spent less time agonising over how to become more individually sustainable (Gibson *et al.* 2011) and more time being politically active and pressing for major environmental changes by social institutions and corporations, then we might achieve global environmental change rather than a myriad of minor adaptive responses.

References

Anon (2010) Towards more sustainable supply chains. *ECOS Magazine* **156**, 18–19.

Barnett C, Cloke P, Clarke N and Malpass A (2005) Consuming ethics: articulating the subjects and spaces of ethical consumption. *Antipode* **37**, 23–45.

Burt S, Jackson T and Sparks L (2002) Retail internationalization and retail failure: issues from the case of Marks and Spencer. *International Review of Retail, Distribution and Consumer Research* **12**, 191–219.

Campbell H (2009) Breaking new ground in food regime theory: corporate environmentalism, ecological feedbacks and the 'food from somewhere' regime. *Agriculture and Human Values* **26**, 309–319.

Dixon J (2002) *The Changing Chicken: Chooks, Cooks and Culinary Culture*. University of New South Wales Press, Sydney.

Dovers S, Hutchinson M, Lindenmayer D, Manning A, Mills F, Perkins P, Sharples J and White I (2008) Uncertainty, complexity and the environment. In *Uncertainty and Risk: Multidisciplinary Perspectives*. (Eds G Bammer and M Smithson) pp. 245–260. Earthscan Publications, London.

Erickson P, Bohle H-G and Stewart B (2010) Vulnerability and resilience of food systems. In *Food Security and Global Environmental Change*. (Eds J Ingram, P Erickson and D Liverman) pp. 67–77. Earthscan Publications, London.

Freidberg S (2004) The ethical complex of corporate food power. *Environment and Planning D* **22**, 513–531.

Gibson C, Head L, Gill N and Waitt G (2011) Climate change and household dynamics: beyond consumption, unbounding sustainability. *Transactions of the Institute of British Geographers* **36**, 3–8.

Hoeschele W (2010) *The Economics of Abundance: A Political Economy of Freedom, Equity and Sustainability*. Gower, Surrey.

Humphery K (1998) *Shelf Life: Supermarkets and the Changing Culture of Consumption.* Cambridge University Press, Cambridge.

Humphery K (2010) *Excess: Anti-consumerism in the West.* Polity Press, Cambridge.

Johnston J, Biro A and MacKendrick N (2009) Lost in the supermarket: the corporate-organic foodscape and the struggle for food democracy. *Antipode* **41**, 509–532.

Jones P, Comfort D and Hillier D (2005) Corporate social responsibility and the UK's top ten retailers. *International Journal of Retail and Distribution Management* **33**, 882–892.

Kagawa F (2007) Dissonance in students' perceptions of sustainable development: implications for curriculum change. *International Journal of Sustainability in Higher Education* **8**, 317–338.

Kjaernes U, Jacobsen E and Dulsrud A (2008) Consumer choice as a mode of governance: the case of farm animal welfare in Europe. National Institute for Consumer Research, Oslo. <www.nsd.uib.no/polsys/data/filer/aarsmeldinger/AN_2008_15601.pdf> (accessed 10.03.10).

Lang T, Barling D and Caraher M (2009) *Food Policy: Integrating Health, Environment and Society.* Oxford University Press, Oxford.

Larsen K, Ryan C and Abraham C (2008) *Sustainable and Secure Food Systems for Victoria: What do we Know? What do we Need to Know?* Victorian Eco-Innovation Lab (VEIL) and Australian Centre for Science, Innovation and Society, University of Melbourne, Melbourne.

Marks and Spencer Group (2010) *Our Plan A Commitments 2010–2015.* <plana.marksandspencer.com/media/pdf/planA-2010.pdf> (accessed 10.05.10).

Mitchell V-W and Kiral H (1999) Risk position of UK grocery multiple retailers. *International Review of Retail, Distribution and Consumer Research* **9**, 17–39.

Mitchell V-W and Harris G (2005) The importance of consumers' perceived risk in retail strategy. *European Journal of Marketing* **39**, 821–837.

Mitchell M, Tregidga H and Walton S (2009) Words not actions! The ideological role of sustainable development reporting. *Accounting, Auditing and Accountability Journal* 22, 1211–1257.

New Economics Foundation (2009) *The Consumption Explosion.* New Economics Foundation, London. <http://www.neweconomics.org/publications/consumption-explosion> (accessed 15.03.10).

SCR (Sustainable Consumption Roundtable) (2006) I will if you will: towards sustainable consumption. Joint initiative of the Sustainable Development Commission and National Consumer Council, London. <www.sd-commission.org.uk/publications/downloads/I_Will_If_You_Will.pdf> (accessed 15.09.09).

Seyfang G (2006) Ecological citizenship and sustainable consumption: examining local organic food networks. *Journal of Rural Studies* **22**, 383–395.

Shove E (2003) *Comfort, Cleanliness and Convenience.* Berg, Oxford.

Shove E, Trentmann F and Wilk R (2009) *Time, Consumption and Everyday Life.* Berg, Oxford.

Sustainability Institute (2003) Commodity systems challenges: moving sustainability into the mainstream of natural resource economies. Sustainability Institute, Hartland. <www.willamette.edu/centers/publicpolicy/projects/oregonsfuture/PDFvol12no1/Sawin1-2> (accessed 01.10.10).

14 CRISIS, CHANGE AND WATER INSTITUTIONS IN SOUTH-EAST QUEENSLAND: STRATEGIES FOR AN INTEGRATED APPROACH

Tabatha Wallington, Cathy Robinson and Brian Head

Introduction

With extended drought conditions and unseasonal floods affecting many regions of the world, uncertainty has become a defining characteristic of water planning and management (Pahl-Wostl 2007). The unpredictability and complexity of water systems and their processes, and the ambiguity that results from multiple understandings of the problem and different normative judgements of acceptable risk, has led to calls for a relational and contextual approach to uncertainty in water management (Brugnach *et al.* 2008; Isendahl *et al.* 2010). This chapter is focused on the uncertainty challenges posed by the recent 'millennium drought' in south-east Queensland (SEQ), located on the subtropical east coast of Australia. Reduced levels of rainfall recorded from 2001 marked the onset of a drought later recognised to be the worst in recorded history (QWC 2008) – a drought of crisis proportions.

The air of crisis surrounding contemporary discourses on water supply and management has posed novel challenges, an influential source of which is the progressive reframing of the water issue itself. When drought is defined in terms of localised and transient conditions of scarcity, as has traditionally been the case, access to water is managed and supply assured through the control and capture of environmental flows – the 'pipes and dams' approach (Ioris 2008). The reliability of rainfall predictions, based on averages calculated over the past 100 or so years, has contributed to confidence in prevailing expert-led approaches to water management. Risks are minimised based on the best available information from hydrological records and are controlled with engineering knowledge of large-scale technologies, which ensures that large margins of safety are built into the water system in the form of large-scale infrastructure. Such confidence is being increasingly undermined, however, as climate variability and population growth have led to a questioning of the capacity of these conventional knowledge bases and institutions to secure a reliable supply of water for future needs.

In SEQ, challenges to the traditional institutions of water management since the onset of drought have had a widespread impact. First, the uncertainties associated with the science that has traditionally informed water planning are now widely recognised, so that established methods of water planning based on historical average rainfall patterns are being revised. Second, predictions of an increasingly variable climate and of more frequent extreme weather events have led to questioning of traditional approaches to management based on prediction and control. Third, the decentralised organisational environment for water planning and supply – where ownership of assets and responsibility for operations rested with 18 local councils – was described as 'parochial and fragmented' (QWC 2008, p. 41) and deemed to need greater coordination in order to optimise the use of scarce water supplies across the region.

In response to these challenges associated with the knowledge base, the management paradigm and the organisation of water management, the Queensland government initiated a comprehensive reform of water institutions in SEQ. We suggest that the focus on institutional reform reflects a deeper uncertainty about the capacity of traditional institutions and their prevailing knowledge cultures to effectively govern water management. In particular, policy and institutional reform associated with the drought reflects a decisive shift in the framing of water issues that is inextricably linked to a shift in the meaning of risk. Where it had been assumed that risks associated with natural hazards such as drought could be minimised through expert knowledge and large-scale infrastructure, risk takes on a more social and political character with the framing of drought as a crisis of governance. Risk management, in turn, demands attention to the institutionalised social relations that govern water practices and regulate uncertainty within the water system.

Insights on risk from social theory, and recent commentary in the water governance literature, both point to the priority of integrating the range of knowledge and perspectives in an open-ended process that engages the range of social actors through more collaborative and dialogue-based institutions. Three decades of social analysis has helped to enrich the concept of risk beyond the probability of harm arising from physical, biological or social causes and accessible only to experts, to recognise its historically, politically and culturally embedded character (Jasanoff 1999; Renn and Schweizer 2009). This has been met with a call for 'technologies of humility' – institutionalised policy strategies that incorporate the expectation of unforeseen consequences and acknowledge the plurality of perspectives and the need for mutual learning (Jasanoff 2002). In the water governance literature, stakeholder participation, institutional collaboration and social learning are increasingly presented as promising means of integrating diverse knowledge and perspectives to facilitate the co-production of knowledge and action under conditions of uncertainty (Lane and Robinson 2009; Pahl-Wostl *et al.* 2010; Saravanan *et al.* 2009; Steyaert and Jiggins 2007).

In this chapter, we adopt a sociological perspective to explore the institutional strategies for integrating knowledge and actors in SEQ water governance, adopted in response to the recent water 'crisis'. We elaborate the value of an institutional perspective on stability and change in water governance regimes, before explaining our methodology and research approach. The sources and effects of uncertainty in the SEQ water system are then described, with a focus on the influence of problem reframing and risk-spreading strategies introduced with governmental reforms, and on the factors limiting the shift to a more integrated approach in SEQ. The final section draws lessons from the SEQ case analysis, and from international experiences, to better understand the phenomenon of crisis response and the institutions that seem more likely to be resilient in the context of social and environmental change.

Understanding institutional stability and change

Institutions are vital determinants of society's capacity to govern natural resources (Mehta *et al.* 1999). They represent 'the multitude of means for holding society together, for giving it a sense of purpose, and for enabling it to adapt' (O'Riordan and Jordan 1999, p. 81). While institutions are generally valued as a relatively stable set of practices and rules that structure human behaviour (March and Olsen 1999), constantly changing social, environmental and technological circumstances generate pressure on society to adapt by means of institutional innovation (Livingston 1993; Saleth and Dinar 2000). The inherent stability of institutions is thus problematic in terms of the desired capacity for flexibility in the face of social and environmental change. This is particularly acute in water governance regimes.

In Australia, institutional analyses of water governance regimes have found that the characteristics of traditional water management arrangements pose significant barriers to change (Livingston *et al.* 2004; Brown 2005). Urban water management has traditionally been dominated by an industrial engineering paradigm characterised by expert knowledge, practices of optimal design under predictable conditions and technical approaches to risk management that seek to quantify uncertainties (Colebatch 2006). The co-evolution of water institutions – the rules, norms, values and shared knowledge of practitioners – alongside large-scale technological infrastructure and stakeholder expectations regarding the reliability and quality of water supply generates an interdependence that makes urban water regimes resistant to change. In other words, inertia in both the physical hardware (longevity of infrastructure and resource commitments) and the institutional software (shared knowledge, rules and habits) act to prevent change (Pahl-Wostl 2009; Sendzimir *et al.* 2010).

The legacy of an expert culture based on engineering knowledge of technological solutions to narrowly defined problems has meant that traditional systems of water management are insulated and risk-adverse (Lach *et al.* 2005). As such, they exhibit limited capacity for integrating different actors and different types of knowledge, a capacity that may be needed to fully understand resource governance problems and to identify innovative solutions to address them (Berkes and Folke 2002). Different water practices are instead managed through specialised institutions, differentiated in terms of functional areas of expertise. This approach is increasingly problematic when responding to changing social and environmental conditions requires the continuous transfer and integration of knowledge across functional and organisational boundaries to facilitate and maintain long-term problem-solving capacity (Ingram and Bradley 2006; van Kerkhoff and Lebel 2006; Lane and Robinson 2009).

Proposals for communication across the fragmented institutional landscape that characterises water management have been dominated by calls for the integration of all aspects of water management, preferably at the regional or catchment level (Senate Committee 2002; Global Water Partnership 2010). The benefits of cooperative and collaborative strategies to facilitate structured interaction between multiple actors, organisations and forms of knowledge in water governance regimes have been highlighted in Australia (van de Meene *et al.* 2009; Robinson *et al.* 2009) and internationally (Steyaert and Jiggins 2007; Pahl-Wostl 2009). Collaboration between government and non-government sectors and other civil society actors aims to manage conflict and enhance social and institutional responsibility, to enhance the coordination of effort required for more efficient and responsive management approaches and to enable the sharing of knowledge and the inclusion of diverse perspectives to inform management programs (Lane and Robinson 2009). Communication and knowledge transfer are vital to such collaborative efforts if advances in scientific and other forms of knowledge are to be translated and integrated into organisational decision-making (Owens *et al.* 2006). The institutional question concerns what kind of relations can be built between different actors (or social roles) and different forms of knowledge to achieve a more integrated approach to water governance (Healey 2008).

The choice of decision-making model is important here, because the various models understand the communication, use and transfer of knowledge in different ways. The desire to optimise performance in the prevailing rational choice model (Jaeger *et al.* 2001) of water management assumes that managers are strongly motivated to incorporate research results into their decision-making. However, while empirical studies show that water managers have a positive attitude toward the use of scientific information in decision-making, they also reveal that such information is rarely acted upon directly (Rayner *et al.* 2005). More generally, the assumption that good information will automatically influence strategic policy and management decisions has been discredited by a growing stock of evidence (Cash *et al.* 2002; Roux *et al.* 2006). Sociological analyses help to explain this situation, finding that the use of information by water managers is inextricably connected with the shared meaning and identity within an organisation as well as with the collective pursuit of implicit organisational goals (Rayner *et al.* 2005). A sociological institutional perspective is therefore warranted to assist in understanding the influence and sources of uncertainty in water management and their effect on the capacity of traditional institutions and their knowledge cultures to effectively govern water systems.

Research approach

Regulating uncertainty: a sociological perspective on institutions

Institutions are overarching systems of beliefs, values, traditions, norms, rules and practices that shape or constrain social behaviour (Pierre 1999). They frame and shape the way participating subjects are brought together and relate to each other and influence the knowledge and resources that are brought to bear on the realm of collective action (Pellizzoni 2004). The analytical framework developed by Scott (1995) articulated the relationship between the cognitive, normative and regulative aspects of institutions. The first 'cognitive' element is characterised by shared understandings or 'problem framings', which influence the kinds of knowledge and expertise considered legitimate for shaping and addressing the problem. The second 'normative' element consists of the shared values and norms, which define roles and responsibilities for action. The third 'regulative' element refers to the rules and sanctions that regulate social interaction in pursuit of shared values. This framework is based on a clear distinction between institutions and organisations (North 1990; Scott 1995); the relationship between the two – between systems of values and norms on one hand, and the organisational system of government on the other – is a dynamic rather than a naturally harmonious one (Pierre 1999).

A fundamental quality of institutions is their effect in reducing uncertainty and stabilising human behaviour (North 1990). The institutionalisation of human behaviour means the emergence of relatively stable, reciprocal expectations between individuals, groups and organisations concerning what are to be regarded as suitable models of action or relationship. In this way, institutions reduce the uncertainty associated with the reciprocal dependency of social relations through rules that restrict the possibilities of agency and choice (Pellizzoni 2004). They also reduce the transaction costs of cooperation that stem from obtaining information about other agents in cooperative relationships and from enforcing agreements made with those agents. Through this structuring of the interactions around resources, institutions aim to reduce uncertainty by promoting the stability of expectations *ex ante* and consistency in actions *ex post* (Agrawal and Gibson 1999).

Viewed as normatively compelling modes of human behaviour, institutions exhibit a constant interplay between their cognitive and normative dimensions: norms focus attention on different aspects of the world, while cognition compels action by providing information on how norms can be fulfilled, modified or discarded (Pellizzoni 2004). Rules, sanctions and rewards then structure the way a social endeavour or practice is organised to pursue shared norms and values. Human interactions with the water cycle are defined and shaped by a range of such practices, which become stable over time, or 'institutionalised' (Colebatch 2006). The condition of institutional inertia or stability is maintained when the institutional framework reduces uncertainty by creating cognitive 'black boxes' – aspects of the problem situation that can be taken for granted (Pellizzoni 2004). By erecting boundaries around uncertainty in this way, for example about the issue or the preferences and abilities of other actors, it can be dealt with through organisational rather than behavioural change. Uncertainty is regulated through rules and sanctions that produce reciprocal expectations. Pressures to conform, such as through coercion or the professional standardisation of techniques, have the effect of reproducing the dominant cooperative scheme (Buchanan 2006). In contrast, opening the cognitive black box can have the effect of challenging the problem-framing by exposing values, norms or beliefs that are not shared, thus disrupting institutional stability. In this way, new knowledge might challenge institutional inertia to catalyse change or competing interpretations of the problem might lead to conflict and catalyse change.

Research methods

A sociological institutional approach, as outlined above, was employed in this study to improve the understanding of institutional stability and change in the SEQ water sector. Investigations focused on the institutional characteristics associated with water reform from around mid 2004 in response to the severe and extended drought. Methods included literature review, document analysis (e.g. policy strategies, Hansard, public submissions) and semi-structured interviews. Interviews were conducted between December 2008 and February 2009 with 15 key informants with extensive experience in water management in SEQ. The interviews were recorded and transcribed, and analysed using Filemaker Pro computer-assisted qualitative data analysis software. The software was utilised to identify the main themes and ideas raised within each interview transcript. This aided in systematising analytical procedures (Miles and Huberman 1994) and assisted the process of storing, searching and managing data. Internal reliability was strengthened through cross-checks on coding and analysis by three researchers. External verification was progressed by inviting a review of the draft report by the study participants to ensure reliability of interview data analysis.

The theoretical framework elaborated above informed the articulation of open-ended interview questions.

1 Framing the problem – how was the water 'problem' defined at the 'moment' of policy change?
2 Strategies used to deal with risk and uncertainty – what strategies were adopted to manage risk and offset uncertainty? What were the roles and responsibilities of different institutions?
3 Limiting factors – which factors influenced the capacity to deal with the 'problem' and limited the adoption of a more integrated response?

The interplay between the cognitive and normative dimensions of institutions, and the consequent influence of this interplay on organisational strategies for integration, is discussed below to explore the sources and effects of uncertainty in the SEQ water system.

Strategies for integration in SEQ

Problem/solution: securing water supply

Until the onset of drought, the water 'problem' in SEQ had been thought of more in terms of having too much water rather than not enough. Policy-makers who remembered the 1974 flood that had inundated Brisbane city and many of its suburbs confirmed that, in the years surrounding the flood, oversupply of water was considered to be the more pressing policy problem. Infrastructure provision in the form of the Wivenhoe dam was quickly commissioned and built along the Brisbane River for the purpose of flood control rather than water storage.

By the mid 1990s, the question of water storage infrastructure was on the radar of water professionals in government agencies, who had turned their attention to water supply planning as a contingency against the impacts of climate variability and population growth on water provision. Later, regional planning exercises for water supply adopted an extended time horizon for forecasting supply and demand, as is typical of inherently conservative and risk-adverse water institutions (Lach *et al.* 2005):

> That's a key part – going right back to 2002, the reason we started the regional water supply strategy is that we wanted to take a 50-year view of what was going to be the demand and supply requirements for south-east Queensland.

As a result of this planning process and in light of the drought conditions being experienced in other Australian capital cities, recommendations to government in 2004 included dam-building and other contingency measures to plan for the possibility of a potentially prolonged period of low rainfall in SEQ (Queensland Government 2004, 2005). This advice was not acted on by government at that time, however. One interviewee described the situation as a water supply planning vacuum in terms of strategic government commitment, a vacuum caused by the legacy of understanding water as an issue of oversupply rather than scarcity. Notwithstanding this lack of policy action, by mid 2004 there was a sense of deepening crisis in SEQ owing to growing media attention to the possibility of a severe and extended drought, given that water levels in major dams had fallen to 50% of capacity. There was a distinct shift to a focus on water scarcity rather than on oversupply of water and flood mitigation.

Emerging uncertainties associated with the knowledge base previously relied upon were key to this problem reframing. In particular, the assumption that historic rainfall records were a reliable basis for future water planning was overturned by the experience of the drought. In turn, uncertainties associated with the scientific basis for risk management prompted greater awareness of climate variability and change and of the risk of surprise events. Combined with falling dam levels, these issues meant that providing a reliable supply of water for the expected population growth in SEQ was an increasingly complex undertaking. Bureaucrats sought to learn from the experience of policy-makers in other Australian cities, where extended drought conditions had led to a strategy of flexibility in policy development and implementation. The orientation of water management institutions increasingly shifted away from the structural approach that had traditionally characteristised water supply provision and toward an iterative risk-based approach to managing water demand at the household level. The need to consider the social dimensions of this strategy – contextual dimensions previously treated as external to the integration of physical aspects of the water system – were not lost on the water professionals we spoke to. As one interviewee noted:

> [We became aware that] you needed to get into restriction regimes to ... pull demand off earlier rather than later ... to give yourself time to work with people.

The nature of the challenge posed by the drought in SEQ was both complex and contested. Water had become a 'wicked problem', characterised by a combination of scientific uncertainty, competing values and disagreement about solutions (Head 2008). According to the still-challenging critique of rational planning (Rittel and Webber 1973), such complex modern problems are generally ill-defined and rely more on political judgement than on scientific certitude. Around 2006 such a judgement call was made:

> And so that's when the policy and the politics and the science kind of merged at that point, and we actually got an understanding that said two things. One is, we haven't been planning on the right basis in south-east Queensland – we now need to learn a new way of planning. And two, it follows from that, that our institutional arrangements to manage this aren't going to cope.

The centrepiece of the government's institutional response to the drought was water security, understood in terms of the security of water supply. This definition of the problem simultaneously brought the solution: when the problem is defined as a supply-side deficit, the response automatically becomes one of enhanced supply-side solutions. A range of solutions was proposed to meet the water security imperative (QWC 2008). New infrastructure took the form of a $AU9 billion water grid to link water across the SEQ region from existing and proposed dams and from new desalination and potable water recycling facilities, with all management responsibilities transferred from local councils to new state government entities. Demand management strategies were also introduced, such as restrictions on household water use and a ban on outdoor hosing that was unprecedented in SEQ. Overall, the new strategy might be considered a precautionary risk-based approach – one that aimed to avoid the risk that water shortage will disrupt the everyday water practices that have come to define the SEQ life-style (QWC 2008).

Sharing responsibility, spreading risk

The policy response to the growing sense of crisis had complex foundations. On the one hand, a range of cooperative and collaborative strategies had been developing over an extended period through the 1980s and 1990s to address interconnected issues of sustainable development and integrated natural resource management (Wallington *et al.* 2010). Cooperative agreements reached by all states and territories at the national level, including the Water Policy Agreement (Australian Government 1994) and the National Water Initiative (Australian Government 2004), had translated into cooperative strategies at the regional level in SEQ. An alliance of the mayors of all SEQ local government councils, the Council of Mayors, was formed to cooperate on integration in the form of planning for total water quality management, which recognised the interrelationships between the human uses of water and its role in the environment. New values around sustainability were addressed through the emergence of a new organisation, the Healthy Waterways Partnership, dedicated to a collaborative approach to deteriorating water quality in SEQ rivers, estuaries and bays. The increasing complexity of water management associated with new agendas of ecosystem health and the allocation of water for environmental uses prompted the inclusion of new actors (environment agencies, farmers, conservation groups and industry peak bodies) and the integration of new expertise in the negotiation of the Queensland *Water Act 2000*. The strategy of affording legitimacy to new players was adopted to avoid the conflict that may have resulted if stakeholders were excluded and to ensure that stakeholders would share responsibility for the outcome based on their contribution to, and consequent ownership of, the policy.

On the other hand, the institutional landscape for water supply had become fragmented. The senior government bureaucrats we spoke to were clear in recognising that institutional arrangements for responsible and accountable water governance were not in place, nor was the strategy of cooperation that had developed in response to the agenda of integrated water management deemed capable of responding to drought:

> We didn't have the right institutional arrangements in place in terms of clear responsibilities and accountabilities. There were the huge transaction costs, with the large number of institutions involved, to be able to get the cooperation and coordination and incentives to do the job that needed to be done.

As the pressures for decisive action to deal with the water crisis increased, some of the key players began to frame the problem as one of fuzzy roles and responsibilities between the state and local bodies. According to one interviewee, there were a lot of 'islands [under] different regimes of governance'. Some of these bodies were well managed, others less so. Above all, responsibilities for water supply were not 'integrated across the region'. Decentralised arrangements for service delivery by local councils had been associated with a proliferation of differing policy positions on levels and standards of service, so coordination was deemed necessary to optimise the use of scarce water supplies and develop an agreed water management strategy (Queensland Government 2006). The potential for conflict across local councils meant that the voluntary model of regional collaboration and cooperation among mayors and others for water planning and supply did not survive the state government's agenda for crisis response from 2006. The urgent goal was to build a governance model that would underpin the water security guarantee pledged by the government. As one observer commented, it was a governance problem rather than an infrastructure problem:

> So the challenge was, while there was an interconnected regional supply, there was no interconnected way of achieving efficiency in that regional supply or achieving cooperation and the best outcomes for the region for that supply.

In 2006, the Queensland government initiated a series of major policy and institutional reforms. The first was to introduce the *Water Amendment Act 2006* that created the Queensland Water Commission, a new organisation with the authority to develop a water security strategy for SEQ. The creation of a new organisation was a prominent coordinating strategy, one that had been previously adopted for regional water governance in Australia with the creation of the Murray-Darling Basin Commission and the Wet Tropics Management Authority. The aim was to internalise the multiple sectoral and jurisdictional dimensions of a problem (Lach *et al.* 2005; Lane and Robinson 2009).

Coordination was also achieved by placing contingency planning on a whole-of-region basis, assisted by a large investment in a water grid with greatly increased capacity to move bulk water between major storages. The aim was to increase certainty around water supply by spreading the risk of water shortage across the region. Risk-based planning favoured diversification of water sources beyond existing catchment-based storages, hence the strategic investments announced in 2007 included a new desalination plant and a new potable recycled water plant. The risk-based approach also involved the introduction of more innovative and non-structural solutions such as demand management to complement the traditional supply-oriented approach and its reliance on large-scale infrastructure. The idea of shared responsibility underpinned the strategy of demand management, for example restrictions on outdoor use of water in residential homes, by enlisting the participation of individuals and households. The focus on external water use imbued this strategy with a public rather than a private function, signalling the role of community responsibility in the response to water scarcity (Gearey and Jeffrey 2003). Water restrictions meant that every citizen was enrolled both in modifying their behaviour to limit outdoor water use, and in policing and thereby regulating the actions of their neighbours.

On the whole, the drive for clarity of responsibilities and accountabilities and for simplification of governance arrangements entailed substantial centralisation. Prior to the restructuring, local authorities had owned about 80% of the water infrastructure in SEQ and each council was responsible for water supply and sewerage functions within its jurisdiction. As the water crisis deepened, the state government acquired most of the water infrastructure assets to consolidate control over the resources required to secure supply, and transferred local government roles to a new range of state-owned entities (Harman and Wallington 2010). A centralised locus of responsibility was implicit in the creation of a water grid, which suggests centrally administered, hierarchical relations of political power. The regional spatial framework created by this grid allows patterns of production and consumption to be aggregated, an approach which aimed to overcome the problem of conflicting policy positions across different local councils. Rather than the voluntary approach that had characterised the early days of water supply planning, policy-makers we spoke to emphasised that, at moments of crisis, there was a perceived need to concentrate political authority in the state to regain central control.

At the same time, the broader agenda for sustainable water supply and use remained in the background as an element of the overall water governance regime. Interviewees noted that strategies of cooperation, collaboration and knowledge-sharing were unlikely to maintain a hold when there was intense conflict over resources and when authoritative and responsible action must be taken in the short term to secure the public good – in this case, the good of water security. The responsibility for strategic decision-making based on government authority was generally perceived to be functionally separate from the role of collaboration in the implementation of those decisions.

Limiting factors

The shared responsibility model of risk response associated with the enlistment of individuals and households in the delivery of demand management strategies reflected a somewhat more social understanding of risk response, in that it involved behavioural rather than organisational change. The model was confined to the implementation of regulation strategies, as the wider engagement of citizens and stakeholders was not extended to the cognitve (re)framing of issues and innovation pathways to be adopted in the new water security planning agenda (Harman and Wallington 2010; Keath and Brown 2009). Alternative public meanings and concerns went beyond a focus on targets for water use; for example, social research on the security of water supply services found that it was the service itself (e.g. cleanliness) and related social obligations, rather than the volume of water *per se*, that structured everyday practices at the household level (Shove 2003). The recognition of such alternative public meanings might serve to open the cognitive black box to reframe the shared value of water security in terms of wider societal understandings, and to catalyse institutional change.

The conservative culture of traditional water management institutions has continued to pose a significant barrier to the wider engagement of multiple perspectives in the SEQ water industry. The culture of corporatised water businesses was described by interviewees as risk-adverse (see also Environment Business Australia 2002), a culture that underpinned an accepted water industry strategy for reducing system complexity and uncertainty – the strategy of reducing risk by delimiting the jurisdiction of water management organisations to reflect narrowly defined purposes (Lach *et al.* 2005). The response of local mayors was illustrative of this strategy: when the drought hit, the first line of defence for local councils to meet organisational goals was to consolidate control over water resources within their service area. The impact of this move was felt in Council of Mayors' deliberations when proposals to share water between councils, made by councils affected by the drought, were deemed unacceptable as a topic of discussion by councils not immediately affected by water shortages.

Conclusion: understanding crisis response

The dynamic nature of urban water reform in SEQ in recent years highlights the possibility of institutional innovation and change in a sector long characterised by stability. The analysis presented in this chapter reveals the interplay between cognitive and normative aspects of strategies for institutional integration. From 2004, public recognition of uncertainty around future rainfall patterns was associated with extensive institutional reform to centralise responsibility for water supply security with the Queensland government and to clearly delineate the organisation of management operations by local councils from the strategic decision-making role of the state (see also Colebatch 2006). The need to develop the adaptive capacity of water governance regimes by spreading the risk of scarcity across the water system with the introduction of a diversity of supply and demand management strategies was also achieved by means of structural reform.

At the moment of crisis and policy change, the cognitive and normative uncertainties associated with the intrusion of values seen to be in conflict with the water security agenda, such as environmental quality, were black-boxed. Uncertainty associated with the decentralised water system and with the potential for conflict among local councils was redressed through organisational restructuring and the consolidation of central control. Uncertainty was effectively regulated through the introduction of new rules and a new organisation (the Queensland Water Commission) to reduce the transaction costs of cooperation by reducing the possibilities of local government agency within the new system of organisational routines. The strategy of demand management, a novel innovation within the Queensland water industry (though it had been adopted several years earlier in other capital cities), proved to be the exception to the general rule of coordination. Demand management sought to generate a sense of shared responsibility for reducing the risk of water scarcity by engaging individuals in new water conservation practices and in regulating the outdoor use of water in line with the new restriction regime. That experience was perhaps indicative of the increased flexibility that might be achieved through more widespread institutional innovation in the SEQ water system.

The results of this research indicate that there are a range of different organisational responses to uncertainty and the associated challenges facing water management. As the influence of different values constrains existing decision systems, new organisational forms, structures and responses are created. Rather than replacing existing management strategies, the emerging responses introduced additional strategies for dealing with uncertainty thus a range of institutionalised practices co-exist in a dynamic process of ebb and flow. This conclusion is in line with the experience of water management elsewhere (Lach *et al.* 2005). A key lesson drawn from a comparative study of 16 cases internationally was that the adoption of innovative governance strategies does not necessarily mean that such strategies will displace centralised regulatory approaches (Meijerink and Huitema 2010). In a similar vein, comparative research conducted in European and non-European contexts found that water governance regimes characterised by a balance of top-down, government-dominated processes, alongside bottom-up processes with strong stakeholder participation and integration, exhibited higher levels of learning than regimes dominated by centralised government control (Pahl-Wostl 2009, 2010). One conclusion from this range of experience is that 'integration of valuable new and old components might well form an element of a transition towards more sustainable water management' (Meijerink and Huitema 2010, p. 4).

The analysis presented in this chapter shows that the idea of integration is open to a differentiated response: while it suggests the instrumental goal of coordination, it might also involve the reframing of problems and solutions through collaborative interactions between stakeholders, as was notable in SEQ with attention to total water cycle management by the Council of Mayors. Understanding the combination of instrumental (coordinative) and potentially transformative (collaborative) modes of governance evident in the recent state government response to the water crisis in SEQ thereby requires an acknowledgement of the Janus-faced tension or duality exhibited by institutions – between their outward 'purity' (stability) and lived 'messiness' (instability) (Shackley *et al.* 1996). Perhaps an outward veil of control conceals a deeper ambivalence about the messiness of collaborative modes of governance and their inevitably uncertain outcomes, but also aims to build on the public values and goals created and progressively institutionalised through collaborative practices of decentralised water planning and water quality management.

The question of how collaborative governance processes might contribute to more resilient water governance regimes in the hybrid of centralised and decentralised arrangements that increasingly form the pattern of institutionalised water management practices is deserving of more attention. Dialogue-based strategies in particular hold the potential to open the cognitive black box that reinforces a normative focus on regulating irregularities in contemporary water systems (Pellizzoni 2004). A cognitive openness to uncertainty in the form of alternative practices and perspectives – and the space this creates for innovative reframing of problems based on alternative versions of 'reality' – is more likely to build water governance regimes that are simultaneously more robust and more resilient in the face of change. While some measure of centralised control may be a necessary coordination strategy to secure the public good, such authority needs to be actively maintained through the engagement of citizens and stakeholders in the framing of issues and innovation strategies as well as in the regulatory strategies designed to implement them (Wallington and Lawrence 2008).

Acknowledgements

The authors gratefully acknowledge financial support provided by the CSIRO Water for a Healthy Country Flagship and the Urban Water Security Research Alliance, a scientific collaboration between the Queensland government, CSIRO, Griffith University and the University of Queensland. We would also like to extend our gratitude to the individuals who kindly shared their time and insights with the project team at interviews. Particular thanks are due to Toni Darbas and Tom Measham for their useful feedback on earlier versions of this chapter.

References

Agrawal A and Gibson CC (1999) Enchantment and disenchantment: the role of community in natural resource conservation. *World Development* **27**, 629–649.

Australian Government (1994) Water reform framework. <http://www.environment.gov.au/water/publications/action/policyframework.html>.

Australian Government (2004) National water initiative. <http://www.nwc.gov.au/www/html/117-national-water-initiative.asp>.

Berkes F and Folke C (2002) Back to the future: ecosystem dynamics and local knowledge. In *Panarchy: Understanding Transformations in Human and Natural Systems* (Eds LH Gunderson and CS Holling) pp. 121–146. Island Press, Washington.

Brown R (2005) Impediments to integrated urban stormwater management: the need for institutional reform. *Environmental Management* **36**, 455–468.

Brugnach M, Dewulf A, Pahl-Wostl C and Taillieu T (2008) Towards a relational concept of uncertainty: about knowing too little, knowing too differently and accepting not to know. *Ecology and Society* **13**(2). <http://www.ecologyandsociety.org/vol13/iss2/art30/>.

Buchanan A (1996) Choosing who will be disabled: genetic intervention and the morality of inclusion. *Social Philosophy and Policy* **13**, 18–46.

Cash D, Clark WC, Alcock F, Dickson N, Eckley N and Jäger JS (2002) Credibility, legitimacy and boundaries: linking research, assessment and decision making. KSG Working Papers Series RWP02-046. <http://ssrn.com/abstract=372280>.

Colebatch HK (2006) Governing the use of water: the institutional context. *Desalination* **187**, 17–27.

Environment Business Australia (2002) Senate Environment Committee Inquiry into the management of Australia's urban water. Submission. 13 May 2002.

Gearey M and Jeffrey P (2003) Concepts of legitimacy within the context of adaptive water management strategies. *Ecological Economics* **60**, 129–137.

Global Water Partnership (2010) GWP today. <http://www.gwpforum.org/servlet/PSP>.

Harman B and Wallington T (2010) Institutional arrangements for water management in south-east Queensland. Technical Report No. 21. Urban Water Security Research Alliance, Brisbane. <http://www.urbanwateralliance.org.au/publications/UWSRA-tr21.pdf>.

Head BW (2008) Wicked problems in public policy. *Public Policy* **3**, 101–118.

Healey P (2008) Knowledge flows, spatial strategy making and the roles of academics. *Environment and Planning C: Government and Policy* **26**, 861–881.

Ingram H and Bradley B (2006) Water sustainability: policy innovation and conditions for adaptive learning. Draft discussion paper prepared for the SMEP Academy, pp. 49–77.

Ioris A (2008) The limits of integrated water resources management: a case study of Brazil's Paraíba do Sul River Basin sustainability. *Science, Practice and Policy* **4**, 4–11.

Isendahl N, Pahl-Wostl C and Dewulf (2010) Using framing parameters to improve handling of uncertainties in water management practice. *Environmental Policy and Governance* **20**, 107–122.

Jaeger CC, Renn O, Rosa EA and Webler T (2001) *Risk, Uncertainty and Rational Action.* Earthscan Publications, London.

Jasanoff S (1999) The songlines of risk. *Environmental Values* **8**, 135–152.

Jasanoff S (2002) Citizens at risk: cultures of modernity in the US and EU. *Science as Culture* **11**, 363–380.

Keath NA and Brown RR (2009) Extreme events: being prepared for the pitfalls with progressing sustainable urban water management. *Water Science and Technology* **59**, 1271–1280.

Lach D, Rayner S and Ingram H (2005) Taming the waters: strategies to domesticate the wicked problems of water resource management. *International Journal of Water* **3**, 1–17.

Lane MB and Robinson CJ (2009) Institutional complexity and environmental management: the challenge of integration and the promise of large-scale collaboration. *Australasian Journal of Environmental Management* **16**, 27–35.

Livingston ML (1993) Normative and positive aspects of institutional economics: the implications for water policy. *Water Resources Research* **29**, 815–821.

Livingston DJ, Stenekes N, Colebatch HK, Ashbolt NJ and Waite TD (2004) Water recycling and decentralised management: the policy and organisational challenges for innovation approaches. Paper presented at the WSUD conference.

March JG and Olsen JP (1999) The institutional dynamics of international political orders. *International Organisation* **52**, 943–969.

Mehta L, Leach M, Newell P, Scoones I, Sivaramakrishan K and Way S (1999) Exploring understandings of institutions and uncertainty: new directions in natural resource management. Discussion Paper No. 372. Institute of Development Studies, University of Essex, Brighton.

Meijerink S and Huitema D (2010) Policy entrepreneurs and change strategies: lessons from sixteen case studies of water transitions around the globe. *Ecology and Society* **15**, 21. <http://www.ecologyandsociety.org/vol15/iss2/art21/>.

Miles MB and Huberman AM (1994) *Qualitative Data Analysis: An Expanded Sourcebook*. Sage, Thousand Oaks.

North D (1990) *Institutions, Institutional Change and Economic Performance*. Cambridge University Press, Cambridge.

O'Riordan T and Jordan A (1999) Institutions, climate change and cultural theory: towards a common analytical framework. *Global Environmental Change* **9**, 81–93.

Owens S, Petts J and Buckley H (2006) Boundary work: knowledge, policy and the urban environment. *Environment and Planning C: Government and Policy* **24**, 633–643.

Pahl-Wostl C (2007) The implications of complexity for integrated resources management. *Environmental Modelling and Software* **22**, 561–569.

Pahl-Wostl C (2009) A conceptual framework for analysing adaptive capacity and multi-level learning processes in resource governance regimes. *Global Environmental Change* **19**, 354–365.

Pahl-Wostl C (2010) Water governance in times of change. *Environmental Science and Policy* **13**, 567–570.

Pahl-Wostl C, Holtz G, Kastens B and Knieper C (2010) Analysing complex water governance regimes: the management and transition framework. *Environmental Science and Policy* **13**, 571–581.

Pellizzoni L (2004) Boundaries of uncertainty: deliberative arrangements as institutions in environment and science-related policy-making. Paper presented at the International Sociology Association RC24 Research Committee conference on Local Institution Building for the Environment: Perspectives from East and West. Gorizia, September 2004.

Pierre J (1999) Models of urban governance: the institutional dimension of urban politics. *Urban Affairs Review* **34**, 372–396.

Queensland Government (2004) 'South-east Queensland regional water supply strategy: stage 1 report'. October 2004. Qld Dept of Natural Resources and SEQ Regional Organisation of Councils, Brisbane. <http://www.seqwaterstrategy.qld.gov.au/projects/seqrwss_stage1.pdf>.

Queensland Government (2005) 'South-east Queensland regional water supply strategy: stage 2 interim report'. November 2005. Qld Dept of Natural Resources and SEQ Council of Mayors, Brisbane. <http://www.seqwaterstrategy.qld.gov.au/projects/seqrwss_stage_2.pdf>.

Queensland Government (2006) *Water for South-east Queensland: A Long-term Solution*. Qld Dept of Natural Resources, Mines and Water, Brisbane.

QWC (Queensland Water Commission) (2008) Water for today, water for tomorrow: south-east Queensland water strategy. Draft. March 2008. <http://www.qwc.qld.gov.au/myfiles/uploads/SEQWS/SEQWaterStrategy-Draft.pdf>.

Rayner S, Lach D and Ingram H (2005) Weather forecasts are for wimps: why water resource managers do not use climate forecasts. *Climatic Change* **69**, 197–227.

Renn O and Schweizer PJ (2009) Inclusive risk governance: concepts and application to environmental policy making. *Environmental Policy and Governance* **19**, 174–185.

Rittel HWJ and Webber MM (1973) Dilemmas in a general theory of planning. *Policy Sciences* **4**, 155–169.

Robinson C, Eberhard R, Wallington T and Lane M (2009) 'Institutional collaboration for effective environmental governance in Australia's Great Barrier Reef'. Technical Report. Water for a Healthy Country, CSIRO, Brisbane. <http://www.clw.csiro.au/publications/waterforahealthycountry/2010/wfhc-GBR-collaborative-decisions.pdf>.

Roux DJ, Rogers KH, Biggs HC, Ashton PJ and Sergeant A (2006) Bridging the science–management divide: moving from unidirectional knowledge transfer to knowledge interfacing and sharing. *Ecology and Society* **11**, 4.

Saleth RM and Dinar A (2000) Institutional changes in global water sector: trends, patterns and implications. *Water Policy* **2**, 175–199.

Saravanan VS, McDonald GT and Mollinga PP (2009) Critical review of integrated water resources management: moving beyond polarised discourse. *Natural Resources Forum* **33**, 76–86.

Schlüter M, Hirsch D and Pahl-Wostl C (2010) Coping with change: responses of the Uzbek water management regime to socio-economic transition and global change. *Environmental Science and Policy* **13**, 620–636.

Scott RW (1995) *Institutions and Organisations*. Sage, Thousand Oaks.

Senate Committee (2002) 'The value of water: inquiry into Australia's management of urban water'. Report of the Senate Environment, Communications, Information Technology and the Arts References Committee. Senate Printing Unit, Parliament House, Canberra.

Sendzimir J, Pahl-Wostl C, Flachner Z and Knieper C (2010) Stalled regime transformation in the upper Tisza river basin: the dynamics of linked action situations. *Environmental Science and Policy* **13**, 604–619.

Shackley S, Wynne B and Waterton C (1996) Imagine complexity: the past, present and future potential of complex thinking. *Futures* **28**, 201–225.

Shove E (2003) *Comfort, Cleanliness and Convenience: The Social Organisation of Normality.* Berg, London.

Steyaert P and Jiggins J (2007) Governance of complex environmental situations through social learning: a synthesis of SLIM's lessons for research, policy and practice. *Environmental Science and Policy* **10**, 575–586.

van de Meene SJ, Brown RR and Farrelly MA (2009) Exploring sustainable urban water governance: a case of institutional capacity. *Water Science and Technology* **59**, 1921–1928.

van Kerkhoff L and Lebel L (2006) Linking knowledge with action for sustainable development. *Annual Review of Environment and Resources* **32**, 1–33.

Wallington T and Lawrence G (2008) Making democracy matter: responsibility and effective environmental governance in regional Australia. *Journal of Rural Studies* **24**, 277–290.

Wallington T, Robinson C and Head B (2010) 'Institutional conditions for sustainable and integrated water management in south-east Queensland: interview results'. Technical Report No. 22. Urban Water Security Research Alliance, Brisbane. <http://www.urbanwateralliance.org.au/publications/UWSRA-tr22.pdf>.

15 USING HOLISTIC SCENARIOS TO REWRITE RURAL FUTURES

Tira Foran and Louis Lebel

Introduction

This chapter reviews three independent applications of holistic rural development scenario work. By 'holistic' scenarios we mean structured sets of narratives that explicitly consider a range of social and ecological forces, in a qualitative and participatory manner. In each case reviewed, scenarios made modest contributions to policy processes. The scenarios reviewed consisted of more- and less-desirable variations around a dominant development pathway, with one scenario in the set conveying a particular 'pro-local' vision of rural development. Scenarios, when used to communicate about alternative and preferred courses of action, considering their explicit and implicit risks, can influence policy change. Development decisions can be informed by generating transparent and well-reasoned counter-narratives, which diverge from dominant story-lines.

Such counter-narratives can help identify robust policy and project-level interventions. In short, participatory scenario-building activities can help inform pro-poor development, reducing risks posed for local communities by particular higher-scale development story-lines, possibly leading to improved legitimacy in decision-making.

After a very brief introduction to Thailand, the setting for the cases presented here, we review the concept of holistic scenarios. A comparative and broader discussion follows the case studies. We discuss issues that should be included in any pro-poor rural scenario activity, as well as process-related choices and trade-offs involved in linking policy audiences with scenario work.

Thailand has been south-east Asia's fourth most prosperous economy (in terms of GDP per capita) since the 1960s, as a result of a historically open economy and large-scale development assistance support by the US during the Indochina War. Based on multi-cropping in the central Chao Phraya basin, it became the world's number one rice exporter. After a shift towards greater economic liberalisation in the 1980s, Thailand's industrial growth and the accompanying movement of people out of agriculture (seasonally or permanently) also quickened (Phongpaichit and Baker 1995). Urbanisation became an important driver and consequence of further industrialisation (Rigg and Nattapoolwat 2001). However, Thailand's economic development involved a number of governance and sustainability challenges, beginning with the failure of agricultural development to eradicate rural poverty (UNDP 2009). Rural households face chronic debt as a result of predominantly low agricultural prices, volatile markets and increased consumption (Thavornthon 2009). Rural households are dependent on non-farm and/or migrant remittance income (Paris *et al.* 2008). A related challenge has been the slow pace of industrialists and industrial institutions in capturing higher-value niches in the global economy, thus perpetuating relatively low-value-added industries (Bello *et al.* 1998). Unequal access to quality health, justice and education exacerbates income poverty and constitutes tangible geographic inequalities. As a result of relatively successful democratisation, discontent with inequality is given voice in a variety of social movements. The high cumulative impact of growth presents further challenge. Greenhouse gas emissions per unit GDP are rising (Nguyen *et al.* 2007). The country relies significantly on crude oil and electricity imports; most of the latter is sourced from hydropower with large increases planned from controversial schemes proposed in the lower Mekong River basin (Molle *et al.* 2009). With water and atmospheric pollution impacts extending beyond national borders, regional organisations such as the Mekong River Commission (MRC) and the Asian Development Bank (ADB) have led programs aimed at harmonising water and energy planning and regulation, although these programs have been constrained by state-centric implementation, modernisation narratives (e.g. the tendency to view fuelwood as an unsustainable and undesirable energy source) and sectoral reductionism (Lebel *et al.* 2010).

In the 21st century, Thailand will face emerging social–economic–ecological tensions. These include the fragility of the political party system, administrative and policy ineffectiveness and ongoing distributional conflicts, both within and across the region's political borders. Thailand also will face exogenous uncertainties, such as protectionism in agricultural export markets, climate change and the threat of major discontinuities in supply of imported fossil fuels. There is no guarantee that in 2111 people will enjoy the same material standards they do today.

Scenario analysis and its uses

Scenario analysis refers to techniques of generating insight about different ways the future might unfold. In the business sector, scenarios have long been used as a tool to explore future risky environments and formulate strategy (van Notten *et al.* 2003; Visser and Chermack 2009). Organisations that work with rural communities have used scenarios to expand the range of alternatives being considered and to get people to think through what might be required to bring about desired futures and avoid undesirable ones (Wollenberg *et al.* 2000). Proponents of scenario methods argue that they can be designed to:

- facilitate social learning, possibly resulting in common understandings, new policy narratives and connections between actors (Da Costa *et al.* 2008);
- empower local people (Wollenberg *et al.* 2000; Lebel *et al.* 2005; Thongbai *et al.* 2006; Enfors *et al.* 2008);
- support public participation (Foran and Lebel 2007; Patel *et al.* 2007; Lebel and Bennett 2008) and embed it in policy-making (Da Costa *et al.* 2008);
- identify robust development interventions at local level (Enfors *et al.* 2008);
- support robust decision-making by policy actors (Chermack 2004; Da Costa *et al.* 2008).

Rural development in changing, open societies is inherently fraught with uncertainty and risk. Aven and Renn (2009) defined risk as uncertainty about the severity of the consequences or outcomes of an activity with respect to something that humans value. Renn and Klinke (Ch. 5, this volume) argued that risk governance needs to create a mutual, dialogic, learning process between risk managers (authorities) and the public. We see holistic scenario-building as a technique with the potential to improve learning around the multiple uncertain aspects of rural development. Participatory scenarios allow the integration of different kinds of knowledge (including experiential and tacit knowledge; see Ch. 5), as well as the identification of a preferred developmental pathway.

For example, prominent scenario processes were organised in South Africa during the final years of apartheid. They included two well-resourced projects led by the private sector in the late 1980s and a university-led multi-stakeholder process (the Mont Fleur process) initiated in 1992 (Galer 2004). The Mont Fleur process included leftist academics and (patiently recruited) members of the liberation movement. Participants agreed that South Africa was facing an interlocking political, economic and social crisis and that simultaneous intervention was needed to avert further deterioration. Participants generated several dozen draft scenarios before narrowing them down into four scenarios for the future of South Africa.

At the organisational level, scenario-building can enhance an organisation's decision-making by uncovering and challenging the assumptions managers use to make decisions. These assumptions are embedded in policy narratives and associated mental models. Existing narratives and models may be inadequate in their treatment of uncertainty and risk, or may be poorly matched to reality (Chermack 2004). With its potential for enabling participatory and reflexive governance, qualitative scenario-building thus appears to be a technique well-suited to the extraordinary demands of modern risk governance (Ch. 5, this volume).

The word 'scenario' is also commonly – and somewhat confusingly – used to refer to what is actually a 'policy alternative' or an 'alternative case.' Thus scenario analysis is often used to refer to modelling or case study analysis. Modelling, however, typically explores the effect of changing the parameter values of a few focal variables.[1] Policy-makers are less familiar with scenario analysis conceived as the systematic exploration of different types of processes across a wide range of uncertainty, including processes from more than one policy sector and level of analysis. Conceived in this way, scenarios are not predictions, forecasts or model runs (Fig. 15.1). Rather, they are plausible stories about how the future might unfold – stories with explicit narrative structure.[2] All policy alternatives represent a different set of risks, implicit or explicit. The goal of formal scenario analysis (also referred to as foresight analysis) is to generate contrasting stories of what the future of a geographical area, a policy sector or an organisation

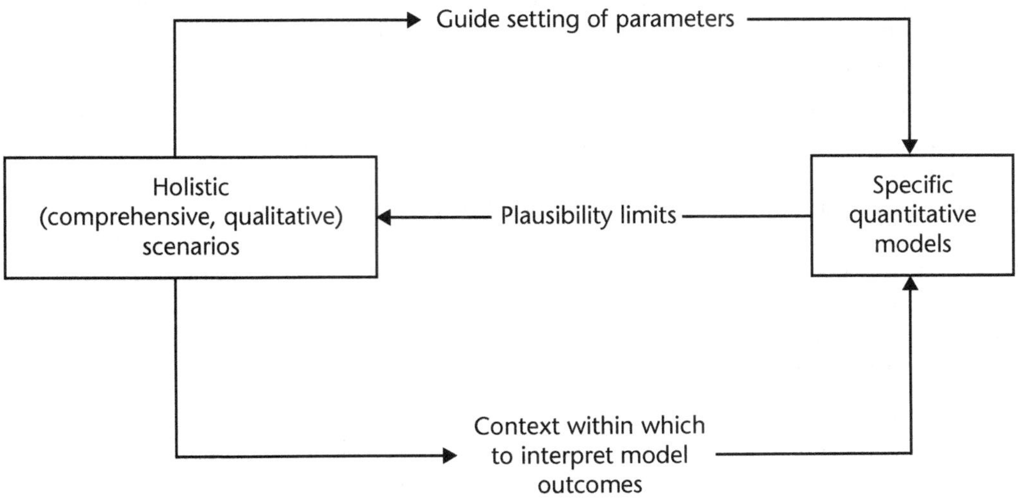

Figure 15.1: Qualitative scenarios complement quantitative modelling.

might look like, depending on plausible combinations of known but uncertain social and environmental forces. The analysis gains insight from the contrast between alternative stories.

Scenarios are holistic to the extent that they attempt to explore the effect of important socio-ecological trends as well as important and uncertain driving forces, across a range of relevant issues. Scenarios processes are effective aids to social learning to the extent that they produce internally coherent stories, allowing for discussion of assumptions and causality (Lebel 2006; Foran and Lebel 2007).

Power of narratives

The notion that scenario-building is a structured form of story-telling directs us to pay attention to how scenarios work as narratives, and how narratives work in policy-making. Scenarios as narratives portray social change, highlighting certain processes, both likely and uncertain, as more interesting and important to the narrators than other processes. The word 'portrayal' suggests oral and written representation as well as visual imagery. Visual imagery can be an effective way to convey the essence of a set of scenarios (Fig. 15.2, colour section).

A development intervention invariably comes with some statement of what the future should be like. Scenarios must somehow link to these policy narratives to be relevant. Hence, to influence and improve governance it is worthwhile thinking about scenarios as candidate policy narratives. Such narratives perform important political tasks: they simplify complex reality through representational and rhetorical devices such as metaphor, analogy, historical references and emotional appeals (Foran 2006). Story-lines (simplified narratives) form the basis of political coalitions. 'Shallow and ambiguous' by character, story-lines allow diverse actors to 'expand their own understanding and discursive competence' (Hajer 1995). They are 'discursive cement that creates communicative networks between actors with different or at best overlapping perceptions' (1995, pp. 62–63). Simplifications allow concrete policy action and problem closure.

For example, Friend *et al.* (2009) argued that in the Mekong region there is an enduring narrative of wild-capture fisheries decline, with four distinctive story-lines:

1 wild-capture fisheries are an open access resource which must inevitably decline in the face of population growth and development;
2 fishing is a marginal activity with limited potential for generating economic development;
3 aquaculture can and should replace wild-capture fisheries;
4 it is necessary to trade-off sustainable wild-capture fisheries for economic development.

The fisheries decline narrative justifies a variety of state-sponsored development interventions such as wetlands reclamation, hydropower development, reservoir construction, aquaculture promotion and fisheries stocking (Bush 2004; Molle *et al.* 2009). Friend *et al.* (2009) critiqued and rebutted each of these story-lines. Regard the second point, they argued that in southern Lao and Cambodia agriculture and fisheries are intertwined and ubiquitous activities, to the extent that it is more accurate to think in terms of river- and wetland-based livelihoods not rice-based livelihoods (2009). Recognising the power of the dominant (but questionable) fisheries decline narrative, Friend *et al.* (2009, p. 325) argued that there is a need to construct a more nuanced and hopeful counter-narrative, or, as they put it, a 'future scenario' of how:

> ... fisheries and the people who depend on them can contribute to setting development objectives ... a future scenario in which fisheries are not merely a resource of conservation value, but a resource whose management is central to meeting the varied developmental challenges of the Mekong River Basin.

This appeal to 'scenario' motivates us to explore whether holistic scenario analysis is a useful method to re-imagine and rewrite the future.

Cases from Thailand

Scenario applications can be classified in terms of their goals, their process design and their content (van Notten *et al.* 2003). Scenario methods and applications are flexible in terms of how they develop each of these dimensions. Tables 15.1–15.3 summarise these features.

Scenario-building multi-stakeholder process in northern Thailand

Chiang Mai province, northern Thailand has been the site of a number of prominent and occasionally violent upstream/downstream conflicts between water users in tributary catchments. In the Mae Chaem watershed, upstream farmers growing maize, soybeans, shallots and other commercial crops have been framed by authorities and downstream users as causing stream pollution and soil degradation. Upstream users have less power than downstream users because of their ethnic minority status and because they may live or farm on land which the national government has claimed as protected areas (Thongbai *et al.* 2006).

Forsyth and Walker (2008) argued that two political coalitions exist, sharing a common belief that Thailand's upland forests are in crisis but differing on the causes of the crisis. A conservative- and state-oriented coalition frames ethnic minority highlanders as forest destroyers because of their slash-and-burn practices. A people-oriented coalition operates with the narrative that highlanders' traditional knowledge is sufficient to return forest ecosystems to health if allowed autonomy and shielded from state regulation and market interventions, including commercial agriculture (Forsyth and Walker 2008). The most obvious dimension of risk is the persistence of a common dominant crisis narrative justifying a range of coercive and risky development interventions, such as involuntary resettlement, reforestation and mandatory land use change.

Scenario-building was part of an ongoing pro-local development intervention called the 'alternatives to slash-and-burn agriculture' project. The scenario application took the form of three specialist-supported scenario-building workshops implemented in the Mae Khong Kha subcatchment, which supports six villages and a population of approximately 3000 people (Thongbai *et al.* 2006). The application can also be considered a multi-stakeholder process in that it involved 25 representatives chosen from local authorities, NGOs and a representative sample of villagers (Table 15.1). Prior to the participatory scenario work, specialists based in northern Thailand conducted a regional-scale scenario analysis (Tomich *et al.* 2005), which Thongbai *et al.* subsequently compared to the results of the local-level scenarios.

Outcomes

Although the stated goals of the project were modest, consisting of exploration and experiment, Thongbai *et al.* reported a breakthrough in cooperative management of water, soil and forest resources at the local level. This began during the second workshop when 'all parties admitted they had equally caused damage' to the watershed. Participants subsequently agreed that one of three scenarios produced was most desirable. This consisted of a localist yet outward-oriented scenario containing 'more alternatives and sustainable, diversified activities':

> S3 ... adopting organic farming or other high-value crops with less land required, using indigenous knowledge and beliefs to conserve forest and natural resources ... reforestation using native plants ... eco- and cultural tourism to generate more

> incomes, and more self-awareness, co-operation and participation ... As a consequence, landscape and natural resources will be restored ... independent livelihoods and well-beings, self-esteem and cultural identity will be sustained (Thongbai *et al.* 2006, p. 8).

Participants formulated a consensus set of recommendations to local authorities. At the end of the process, according to the organisers, participants reported an increase in problem-solving capacity and a sense of empowerment along with recognition of higher-level barriers that needed to be addressed, such as the lack of legal recognition of community forestry (Thongbai *et al.* 2006; Table 15.1).

Scenarios for Pak Mun Dam's operations

Pak Mun Dam, a 300 m long run-of-river dam constructed in 1990–94, is one of Thailand's most controversial infrastructure projects (Foran and Manorom 2009). In 2001, in an attempt to resolve a long-running conflict between hydropower production and fisheries impacts from the dam, the Thai government ordered the dam's operator, the Electricity Generating Authority of Thailand (EGAT) to open all eight sluice gates for four months, May–August. It also mandated the local university, Ubon Ratchathani University, to conduct a multi-disciplinary study on the impacts of opening the gates (UBU 2002). This study was one of several exercises in knowledge production, all of which unfolded in a highly contested setting (Foran and Manorom 2009). The UBU research team presented findings in September 2002, highlighting four management alternatives:

1 maintain the status quo (maximise electricity production; sluice gates normally closed);
2 have a five-month seasonal opening;
3 have an eight-month opening;
4 have year-round opening for five years (no electricity production).

We can consider UBU's four alternatives as a set of specialist-led, decision support, normative scenarios. The structure – a linear examination of the strengths and problems of the first three options, before concluding with the last – suggest that the authors preferred the five-year opening option:

> Alternative 4. Opening the sluice gates year-round. This Alternative follows from the consideration that many technical solutions to the Dam's problems in electricity generation can be found, by obtaining auxiliary energy from other sources in the country and/or importing energy from Lao PDR. However, the problems of community economics do not quite have solutions and are not solvable by technical means. The Dam has affected poverty and natural ecosystem fertility in a manner not possible to deny. Problem solving for more than 8000 households ... will require the resources, cooperation and time of many concerned parties ... Even if the Dam does not generate any electricity, stability of electric current is not negatively impacted. Economic growth, especially growth from industrial development ... has not yet developed as forecasted. The Dam does not yet play a full potential role in irrigation. It is appropriate to direct benefits from the Mun River basin to community-based economics by ceasing use of the Dam for electricity generation ... until electricity demand changes from current conditions. This period ... will provide benefits of restoring ecological conditions, economics and community livelihood. Villagers will be able to access fishery resources throughout the seasonal round ... [increasing] confidence to

Table 15.1: Mae Chaem scenario-building multi-stakeholder process

Goal of scenario work (exploration vs decision support)	**Exploration**
Inclusion of norms	Yes
Vantage point (forecasting vs backcasting)	Forecasting
Subject (issue-, area- or institution-based)	Area and issue (livelihoods, government forest policy, land tenure)
Time scale	Medium-term (20 years)
Spatial scale	Local
Process design (intuitive vs formal analytical)	Intuitive because participatory (specialist scenario workshop followed by 3 participatory workshops (n=25) well distributed in categories such as gender, age, sector, upstream/downstream user)
Participation: mode/level	Direct/high (multi-stakeholder process)
Data (qualitative vs quantitative)	Mixed
Data collection method (participatory vs desk)	Mixed
Resources (limited vs extensive)	Extensive (multiple workshop events supported by national specialists)
Scenario content	
Variables: sector	Rural development
Variables: thematic	Sustainable smallholder agriculture, education and diversification, cultural identity
Variables: actors	Small farmers and state agents
Dynamics (events and processes [drivers])	Increasing household expenses, declining natural resource base, changing regulatory regimes
Temporal nature (sequential or snapshot)	Snapshot
Range of possible futures	Moderate (business-as-usual, agrarian decline, agrarian renewal)
Level of interaction between variables	Moderate
Risk dimensions	Poverty entrenchment Failure to represent legitimate minority interests
Outcomes (stated goals)	
Introduce concept and scenarios process	Yes
Explore participatory NRM methodologies	Yes
Validate the method	Yes
Other outcomes (evident and inferred)	
Changes in positions (discourses)	Yes
Changes in policy narratives	Yes
New agreements between disputants	Yes
Empowerment	Yes
Support for future use of method	Not known (would require post-project engagement with new village leaders)

Source: Thongbai *et al.* (2006); authors' analysis.
First column refers to choices in a typology, adopted from (van Notten *et al.* 2003). 'Change in policy narrative' includes change in understanding and social learning.

> invest both capital and labor [in fishing]. Restoration of nature will occur, as well as restoration of community livelihoods founded in their original locations. Initially, opening the sluice gates year-round might occur for a period of approximately five years ... During this time, there should be preparation to obtain irrigation benefits ... by thorough and detailed surveying and study. Consideration needs to be given to ... building basic infrastructure, policy-making regarding agricultural crops grown in irrigated areas, and cooperative management, the capacity for which many communities already have (UBU 2002, *Khor* 17–18).

The UBU scenarios were generated by specialist interpretation of a large and complex study, drawing in particular on findings about community social relations, local farming systems, fisheries incomes, fisheries catch per unit effort during the experimental dam opening, electric power flow analysis and a stakeholder consultation organised with local authorities. The scenarios attempt to support decision-making in that they focus on one key variable (dam opening and its impacts) while other uncertain drivers (e.g. the sustainability of smallholder farming) are treated in less detail. The scenarios contain the normative story-line that technical substitutes exist for goods provided by the dam, but none exist for improving the security of community-based livelihoods. The explicit prioritisation of community-based livelihoods over energy security leads to the conclusion that any significant opening is preferable to the status quo.

Outcomes

Normative scenarios were an effective way to synthesise and communicate the key findings of a long study. The UBU scenarios succinctly captured a spectrum of strategies with respect to Pak Mun's future operations, including state investment in integrated agriculture and irrigation (Table 15.2). Some critics complained that the alternative scenarios were confusing and did not offer a single explicit set of policy recommendations. It is important to note that by 2002, stakeholders had long aligned into two discourse coalitions holding opposite policy narratives on the value of Pak Mun Dam (Foran 2006). During the history of the conflict, three policy narratives on the future of Pak Mun Dam circulated:

1. beginning 1991 – the dam should be built and continue to operate, but additional efforts at reservoir stocking and livelihood restoration are needed;
2. beginning 1999 – decommissioning is needed to restore migrating fisheries;
3. beginning 2002 – a four-month wet season opening is sufficient to allow fisheries migration and to allow dry season water storage.

Two heterogeneous groups of actors deployed these narratives: and thus different actors used UBU's scenario narratives to justify prior positions. The hydropower project's most vocal supporters rejected the scenarios, accusing the authors of outright bias towards anti-dam activists. The dam's opponents, including some legislators and human rights commissioners, seized on the finding that the dam was not essential for electricity needs and used that as a debating point with EGAT management. During a 2002 seminar organised by the Senate Committee for Public Participation, an admission by the EGAT governor that lower north-east Thailand's electricity demand could be reliably met by imports led the university president to argue for a year-round opening, in a special hearing chaired by Prime Minister Thaksin Shinawatra (Foran 2006).

Actors aligned with the mainstream discourse coalition could refer to several other studies, as well as common-sense lay discourse, to formulate their positions (Foran 2006; Foran and

Table 15.2: Pak Mun scenario application

Goal of scenario work (exploration vs decision support)	**Decision support**
Inclusion of norms	Yes
Vantage point (forecasting v. backcasting)	Forecasting
Subject (issue-, area-, or institution-based)	Issue and area (livelihoods, multi-purpose dam operations)
Time scale	Short-term (5 years)
Spatial scale	Local
Process design (intuitive vs formal analytical)	Not known
Participation: mode/level	Indirect (specialist-conducted)/medium
Data (qualitative vs quantitative)	Mixed
Data collection method (participatory vs desk)	Mixed
Resources for scenario work (limited vs extensive)	Extensive (multi-disciplinary study + 1 synthesis workshop)
Scenario content	
Variables: sector	Rural development
Variables: thematic	Community-based livelihoods, fisheries, irrigation, value of hydropower, migration
Variables: actors	Small farmers, electricity utility, state agents responsible for agricultural development
Dynamics (events and processes [drivers])	Declining natural resource base, ecological restoration, agricultural intensification, changing dam operation regimes
Temporal nature (sequential or snapshot)	Snapshot
Range of possible futures	Moderate (business-as-usual, agrarian decline, agrarian renewal)
Level of interaction between variables	Moderate
Risk dimensions	Failure to sustain living aquatic resources, a common property resource Failure to represent legitimate minority interests
Outcomes (stated goals)	
Decision support	Yes
Other outcomes	
Changes in individual positions (discourses)	Yes
Changes in policy narratives	Yes
New agreements between disputants	Yes
Empowerment	Limited (2002 decision imposed by force)
Support for future use of scenarios method	Not known (ongoing negotiations, politicised context)

Source: Foran (2006); authors' analysis.

Manorom 2009). But among the rival studies and associated policy narratives, the UBU study was unique in that it presented one crystallised set of alternatives which took prior critical narratives seriously. In summary, the politicised arena significantly limited the impact of the UBU study and its synthetic scenario analysis but it did not prevent the main messages of that analysis from reaching decision-makers, including those whose positions were subject to change.

Environmental flows scenarios for the Nam Songkhram basin

In north-east Thailand's Nam Songkhram River basin, an environmental flows study conducted by the Mekong Wetlands Biodiversity Program included scenario-building in its methodology. The extensive floodplain of the lower Nam Songkhram is noted for its high biodiversity, flood pulse ecosystem and wetland-dependent livelihoods. The Nam Songkhram has not experienced the high-intensity conflicts of the lower Mun basin but official plans since the 1980s to build dams and weirs throughout the basin and to build large structures in the lower basin have been challenged as inappropriate by environmental NGOs and by environmental arms of the state. However, they have persisted as orthodox policy narratives. Meanwhile, smallholder farmers have expanded rice and eucalyptus onto the floodplain. Some agribusiness operations, aided by intensification policies, privatised large tracts of floodplain common property, with mixed commercial success (Blake *et al.* 2009).

The Nam Songkhram lacks an integrating river basin organisation. Instead, the basin falls under the jurisdiction of four provincial governments and six subbasin management committees (Blake *et al.* 2010; Blake n.d. b). During the 2000s interest grew in using environmental flows assessment and planning as a way to implement principles of an integrated water resources management model. Environmental flow has been promoted as a participatory, deliberative, ecosystem-based approach to integrated water resources management (Dyson *et al.* 2003), which otherwise might continue to be dominated by economic and engineering approaches. The Nam Songkhram design consisted of specialist multi-disciplinary assessment and scenario analysis followed by a multi-stakeholder dialogue (Table 15.3).

Specialist scenario workshop

The environmental flows team consisted of specialists in hydrology, fisheries, botany and agriculture. As an input to the specialist workshop, four qualitative development scenarios were prepared by the project coordinator:

1 scenario 1 – building two controversial floodplain dams, the Nam Songkhram and the Nam Oon;
2 scenario 2 – expanding irrigation distribution (water grid), i.e. Lao–Thai water transfers, but not the S1 floodplain dams;
3 scenario 3 – business-as-usual, i.e. agricultural intensification, many small infrastructure projects, no major projects;
4 scenario 4 – implementing an alternative development paradigm, i.e. shift to sustainable (low external input) agriculture, focus on demand-driven water resource development, shift to informed, bottom-up, precautionary decision-making, shift to internalising impacts of rural development (after Blake *et al.* 2010).

The first three scenarios shared the dominant (mainstream) story-line that rural poverty alleviation required additional irrigation infrastructure. The first two scenarios were presented in relatively simple terms, consisting of different levels of one focal variable, water infrastructure development (irrigation and flood control structures). The latter two scenarios were more complex. Business-as-usual incorporated trends since the 1970s, assuming continuation of those trends (Table 15.3). Scenario 4 assumed a reversal of those trends, implementing a new and genuinely participatory river basin organisation receptive to existing policy narratives about integrated agriculture (Blake *et al.* 2010).

Table 15.3: Nam Songkhram scenario application

Goal of scenario work (exploration vs decision support)	**Both**
Inclusion of norms	Yes
Vantage point (forecasting vs backcasting)	Forecasting
Subject (issue-, area- or institution-based)	Issue-based (livelihoods, irrigation development, wetland ecosystems)
Time scale	20 years
Spatial scale	Local (3 village sites) and subregional (lower Nam Songkhram basin)
Process design	
Intuitive vs formal analytical	Intuitive
Participation: mode/level	Indirect (specialist-conducted)/medium
Data (qualitative vs quantitative)	Mixed
Data collection method (participatory vs desk)	Mixed
Resources for scenario work (limited vs extensive)	Extensive
Scenario content	
Variables: sector	Rural development, wetlands ecosystem management
Variables: thematic	Water infrastructure, wetland habitat, goods and services, irrigation, community-based livelihoods
Variables: actors	Small farmers, irrigation and agricultural development agencies, politicians, agribusiness
Dynamics (events and processes [drivers])	Loss of upland and wetland forests, agriculture and fishing intensification, wetlands conversion, ecological restoration, decisions to invest in water infrastructure, altered flow regimes
Temporal nature (sequential or snapshot)	Mixed
Range of possible futures	Moderate (business-as-usual, ecological decline, ecological renewal)
Level of interaction between variables	Moderate[a]
Risk dimensions	Failure of infrastructure solutions to control flooding or deliver irrigation benefits Failure to sustain living aquatic resources, a common property resource
Outcomes/effectiveness (stated goals)	
Capacity building (in E-Flows methodology)	Yes
Knowledge production	Yes
Multi-stakeholder dialogue	Yes
Outcomes/effectiveness (other outcomes)[b]	
Changes in individual positions (discourses)	No (not tangibly)
Changes in policy narratives	No (process too short and too restricted in scope to influence)
New agreements between disputants	No (main disputants not present, no follow-up project)
Empowerment	Very limited
Support for future use of scenarios method	Yes (among civil society and progressive state actors)

a: Cross-sectoral interactions recognised but not fully modelled.
b: DJH Blake (*pers. comm.* 24 August 2011).

The qualitative scenarios were fed into a specialist workshop that aimed to describe how each would affect the river flow regime, including events ranging from dry season low flows to large interannual floods.

Scenarios 1–3 all resulted in significant negative ecological impacts. Interestingly, the lower basin floodplain dams scenario (scenario 1) had the largest impact on the hydrological flow events, rather than the Thai water grid scenario. Discussion of social changes focused on probable negative impacts, which ranged from loss of floodplain food and livelihoods and declines in water quality (under business-as-usual), to forced resettlement (scenario 1), to a shift to high-capital, high-risk farming systems with attendant landholding consolidation (scenario 2). Time limitations prevented full analysis of ecological and social impacts, or discussion of scenario 4 (Blake n.d. a).

Multi-stakeholder dialogue

Key findings from the E-Flows assessment and specialist scenario workshop were presented to participants at a day-long facilitated dialogue in 2007. Participants included representatives from national, provincial and local governments, academia, civil society and local communities. The meeting included a two-hour small group qualitative scenario-building activity. At the beginning of the day, participants were given a briefing paper, summarising scenarios 1–4 from the specialist workshop. During the small group activity, they were invited to review scenarios 1 and 3, both scenarios of wetlands loss and agricultural intensification. Participants were asked to discuss 'the current situation, including problems in different dimensions, and to offer clear approaches to solve or mitigate those problems' (Blake n.d. b). Small groups reported back to plenary, sharing and revising sustainable development vision statements, which emerged as counter-narratives to the intensification scenarios discussed. Some of the sketches were highly idealistic with respect to sustaining ecosystem-based livelihoods. However, they emerged from lengthy preceding discussion of the basin's multiple development trends, and can be considered plausible.

Outcomes

During the dialogue meeting, the use of scenarios helped participants focus on floodplain wetlands (ecosystems that are important to poor peoples' livelihoods) and threats to those ecosystems and associated values. Decision support was not a stated goal of the specialist workshop but the focused definition of scenarios 1 and 2, combined with hydrological modelling findings, allowed specific policy arguments to be made, for example that 'any attempt at upstream regulation of the Nam Songkhram or tributaries will have a negligible impact on flood control' (Blake n.d., p. 28). Decision support was a goal, however, of the subsequent dialogue meeting. The meeting discussed multiple management suggestions on topics ranging from livelihoods to conservation to administrative reform. It concluded with discussion of a draft statement calling for the establishment of a four-province interprovincial working group aimed at delivering more integrated basin management. These were initially promising outcomes, but other outcomes were limited by lack of resources with which proponents could follow up (DJH Blake, *pers. comm.* 24 August 2011).

Comparative assessment

The three projects discussed above all involved future scenarios of rural livelihoods in specific localities in the Mekong region. Building normative scenarios was an integral activity in all three. In terms of context, in all three applications, scenario analysis was a component in a larger action research project aimed at improving forest or water resource management in contested settings. In terms of definitions, the Thongbai *et al.* (2006) application was the most sophisticated. For example, the activity was explicitly and correctly identified as a 'visioning' activity (Evans *et al.* 2006). The compatibility of the local-level scenarios with higher-level drivers was explored in a separate step by specialists (Lebel 2006). In terms of relevance, all three studies understood scenarios to be directly relevant in supporting bottom-up participatory planning.

Interestingly, none of the case studies solely used formal scenario methodology (structured exploration of combinations of trends and uncertain social and environmental forces). Rather, the scenarios were developed intuitively, either because the authors regarded their development as self-evident (Pak Mun, Nam Songkhram) or as a deliberate participatory technique (Mae Chaem).

Another finding relates to how the scenarios were organised and presented in textual form. In all case studies, the sets of scenarios produced consisted of more- and less-desirable variations around the status quo, with the final scenario in the set reading as the preferred one. In each case, in the final scenario, small farmers and their locally based livelihoods are sustained by balanced commercialisation and intensification (tourism, high-value-added agriculture, demand-driven irrigation). Such livelihoods and cultures are also sustained by significant local empowerment and improvements in human capital. The final scenario functioned as a detailed policy narrative conveying a vision of rural development. Organised and presented in this way, the exercise helped legitimise a localist counter-narrative (Parnwell 2006).

Political influence

Qualitative and multi-stakeholder scenario exercises for rural futures occur in a variety of settings (Evans *et al.* 2006) but are still rare in Thailand and, based on literature searches, elsewhere in mainland south-east Asia.

Framing the future

Scenario-building involves directly framing future states, impacts and options, including those that are likely, sought and feared. Framing/counter-framing refers to rhetorical action[3] deployed to further a political objective. The discourse in framing/counter-framing can range from terse speech acts (fragments of narratives) to the more elaborate knowledge claims in policy narratives and scientific reports. Framing/counter-framing is a political driver, a social process that contributes to political change (Snow *et al.* 1986; Foran 2006).

The appeal of qualitative methods is that knowledgeable participants or analysts can create complex and plausible narratives, as shown in scenario 3 from Mae Chaem and scenario 4 from Nam Songkhram. The strength of these narratives is their accessibility and rhetorical appeal as visions of sustainable localist development. These eco-localist visions motivate many civil (and some state) actors to implement community-based development, and are especially resonant in Thai democratic politics.

Improving governance

The ability of any discrete method, including holistic scenario planning, to influence governance is always mediated by other drivers of policy processes, including competing narratives, competing knowledge brokers and associated interests and resources (Foran 2006; Foran and Manorom 2009). In all three case studies, use of the tool led to the development of explicit policy recommendations for the increased allocation of resources to address the needs of marginalised smallholders and the reframing of top-down approaches to common property resources. In two cases (Mae Chaem and Nam Songkhram), the recommendations were legitimised by a multi-stakeholder endorsement. In those cases, scenarios played a role in planning processes that were not solely distributive bargaining negotiations.

Generally, natural resource and rural development governance in Thailand has consisted of a series of policy domains dominated by elite actors, institutions and mainstream (modernist) development narratives. Improving natural resource governance in this context involves a host of reforms at different levels. The specific potential of scenario methods is that they offer a means to critique dominant development narratives which, as discourse, constitute powerful contextual drivers of action while at the same time suppressing or normalising particular risks.

As sets of contrasting policy narratives, scenarios highlight the fact that the future is uncertain and open. In the dominant context, scenario-building might be received by policy elites as rhetorical action with the potential to undermine dominant policy narratives (Foran 2006; Snow *et al.* 1986). The more radical or paradigm-breaking the range of plausible futures considered, the less likely such an activity is to offer anything that actors focused on short-term decision-making perceive to be useful. (The need for sustainability transition may be urgent and credible to some, but not to all.) Scenarios by definition present a range of alternate futures, leading to a range of possible value-laden policy narratives. Scenarios designed to offer a reservoir of knowledge for the public might increase the range of potentially receptive audiences. A well-designed communication strategy, making use of audio and visual media, deserves consideration (Da Costa *et al.* 2008).

Scenario analyses complement multi-stakeholder processes, designed to facilitate active and informed participation in particular policy issues (Foran 2010). All such processes include both competitive and cooperative interactions. They may be designed to seek common understandings of a problem, or they may not (Warner 2007; Foran 2010). A multi-stakeholder dialogue can be designed around scenario-building activities (Foran and Lebel 2007). However, if actors join a multi-stakeholder process primarily to engage in distributive bargaining, they are unlikely to be interested in scenario-building or scenario analysis. In the latter situation, scenarios could influence policy processes indirectly, as tools used by some to develop logically robust policy arguments (e.g. rebutting a policy narrative that regional connectivity is fundamental to building prosperity; Foran and Lebel 2007).

Two weaknesses of multi-stakeholder processes are, first, that they can be resource-intensive (all three cases were initiated by research or development actors with limited resources for ongoing engagement), and second that space for participatory processes is limited in non-democratic and democratising countries. Specialist-generated scenarios might mitigate both resource and participation problems. However, specialists eventually need to find receptive audiences, suggesting the need to actively reach out to various policy actors (whether state or civil) using appropriate communication strategies. Specialist scenario organisers need to think about the credibility, salience and timing of their messages, as well as the advantages and disadvantages of having other actors (regional media, NGOs) convey those messages (Da Costa *et al.* 2008). Overall, an iterative scenario process involving specialists, multiple stakeholders and carefully designed opportunities for elites may be most effective.

In a context dominated by conventional development story-lines, scenario activities focusing on rural futures should offer policy or project-level recommendations that would perform well across the range of uncertain futures (Enfors *et al.* 2008).[4] They should be specific enough to inspire and mobilise practitioners.

Despite the generic challenges discussed above, scenario analysis is a method well-suited to exploring, clarifying and rethinking the future of small farmers in Thailand and in other dynamic and contested regions. More holistic scenario applications are needed, combining both specialist and lay stakeholder input.

Conclusion

Water-related decision-making in mainland south-east Asia has been dominated by a number of potent, but simplistic, policy narratives (Friend *et al.* 2009; Foran and Manorom 2009). In these narratives or story-lines, poor rural people can be lifted out of poverty by large-scale water resources development which will make their farmland more productive; likewise, the introduction of modern inputs and practices will make their agriculture more productive. In parallel, the integration of subsistence production into larger-scale markets will also benefit the poor. Vulnerabilities to changing market conditions can be managed through more sophisticated interventions.

This chapter showed how participatory scenario-building can inform pro-poor development, reducing risks embedded in non-reflexive development narratives, thus improving political representation and legitimacy. Because most of the world's poor people still make a living in rural areas (Hazell *et al.* 2007) and because rapid development carries multiple risks and uncertainties (Molle *et al.* 2009), counter-narratives of agricultural and rural change matter. Particularly useful are the narratives which confront uncertainty and risk explicitly, before sketching desired visions. We focused on qualitative and holistic scenario-building as a method of delivering savvy community-based, narrated visions of development. Such applications are still limited in developing countries. In the three independent cases reviewed, holistic scenario methods dealt with a range of uncertainties affecting the future of small farmers in given localities. They produced new local area-based and issue-based policy narratives which could be used in policy advocacy. In all cases, the sets of scenarios produced consisted of more- and less-desirable variations around the status quo, and the final scenario in each set conveyed a desired vision of rural development. In each case, small farmers and their locally based livelihoods are sustained by balanced commercialisation and intensification (tourism, high-value-added agriculture, demand-driven irrigation). Such livelihoods and cultures are also sustained by significant local empowerment and improvements in human capital. These visions underscore the need for reforms that increase downward accountability and effective delivery of resources. Actually attaining a future resembling the one envisioned requires a set of robust interventions and enabling policies at different levels, from local to supranational. Scenarios relevant to marginalised and vulnerable people would have to capture key uncertainties with respect to the future of their livelihood strategies. Many relevant driving forces are determined at levels higher than a specific locality. Scenario analysis ideally would also explore national, regional and global drivers (Lebel 2006; Foran and Lebel 2007), although few examples of multi-level analysis exist (Kok *et al.* 2007; Enfors *et al.* 2008). Further applications of scenario method could attempt multi-level analysis, with analysis of global and regional drivers facilitated by specialists, resulting in a product that combines multiple forms of knowledge (Ch. 5, this volume).

Endnotes

1 In south-east Asia, examples of modelling and case analysis which use the term 'scenario' include a strategic environmental assessment of Vietnam's hydropower development plans (Soussan and Nilsson 2009) and hydrological scenarios commissioned as part of the MRC

basin development planning process (World Bank 2004; World Bank and Asian Development Bank 2006).
2 Models have a narrative structure, but that structure is usually implicit.
3 'Rhetorical action' refers to the competitive, strategic, monologic use of language by actors to persuade audiences of preferred courses of action (Naurin 2007).
4 The argument made here is that the policy recommendation, not the scenario, should be robust.

Acknowledgements

The author thanks Francois Molle, David Blake, Rachel Williams and an anonymous reviewer for detailed and helpful comments on earlier drafts of this chapter.

References

Aven T and Renn O (2009) On risk defined as an event where the outcome is uncertain. *Journal of Risk Research* **12**, 1–11.

Bello W, Cunningham S and Poh LK (1998) *A Siamese Tragedy: Development and Disintegration in Modern Thailand*. White Lotus, Bangkok.

Blake DJH (no date a) Nam Songkhram River Basin E-Flows report: section 4. Nam Songkhram scenarios workshop. Unpublished manuscript.

Blake DJH (no date b) Nam Songkhram River Basin E-Flows report: section 5. Multi-stakeholder dialogue. Unpublished manuscript.

Blake DJH, Friend R and Promphakping B (2009) The Nam Songkhram River Basin landscape transformations and new approaches to wetlands management. In *Contested Waterscapes in the Mekong Region: Hydropower, Livelihoods and Governance.* (Eds F Molle, T Foran and M Käkönen) pp. 173–202. Earthscan Publications, London.

Blake DJH, Sunthornratana U, Promphakping B, Buaphuan S, Sarkkula J, Kummu M, Ta-oun M, Waleetorncheepsawat P, Boonyothayan S, Tharme R, Osbeck M and Janprasart S (2010) E-Flows in the Nam Songkhram River Basin. M-POWER Research Network, Vientiane, Laos. <http://www.mpowernetwork.org/Knowledge_Bank/Key_Reports/Research_Reports/E-Flows_in_the_Nam_Songkhram_River_Basin.html> (accessed 30.09.11).

Bush SR (2004) A political ecology of living aquatic resources in Lao PDR. PhD thesis. Division of Geography, Department of Geosciences, University of Sydney, Sydney.

Chermack TJ (2004) Improving decision-making with scenario planning. *Futures* **36**, 295–309.

Da Costa O, Warnke P, Cagnin C and Scapolo F (2008) The impact of foresight on policy-making: insights from the FORLEARN mutual learning process. *Technology Analysis and Strategic Management* **20**, 369–387.

Dyson M, Bergkamp G and Scanlon J (Eds) (2003) *Flow: The Essentials of Environmental Flows*. IUCN, Gland, Switzerland.

Enfors EI, Gordon LJ, Peterson GD and Bossio GD (2008) Making investments in dryland development work: participatory scenario planning in the Makanya catchment, Tanzania. *Ecology and Society* **13**, 42. <http://www.ecologyandsociety.org/vol13/iss2/art42/>.

Evans K, Velarde SJ, Prieto R, Rao SN, Sertzen S, Davila K, Cronkleton P and de Jong W (2006) Field guide to the future: four ways for communities to think ahead. Center for

International Forestry Research (CIFOR), Nairobi. <http://www.asb.cgiar.org/PDFwebdocs/Evans-et-al-2006-Field-guide-to-the-future.pdf>.

Foran T (2006) Rivers of contention: Pak Mun Dam, electricity planning and state–society relations in Thailand, 1932–2004. PhD thesis. Division of Geography, Department of Geosciences, University of Sydney, Sydney. <http://hdl.handle.net/2123/1984>.

Foran T (2010) Contentious decision-making around Pak Mun Dam. CPWF PN67 case study. M-POWER Working Paper PN67-2010-08. CMU-USER, Chiang Mai.

Foran T and Lebel L (2007) 'Informed and fair? Water and trade futures in the border regions of mainland southeast Asia'. Report from an M-POWER Dialogue workshop. Draft for public comment. USER Working Paper WP-2007-14. Unit for Social and Environmental Research, Chiang Mai University, Chiang Mai. <http://www.mpowernet.org/mweb.php?pg=81>.

Foran T and Manorom K (2009) Pak Mun Dam: perpetually contested? In *Contested Waterscapes in the Mekong Region: Hydropower, Livelihoods and Governance.* (Eds F Molle, T Foran and M Käkönen) pp. 55–80. Earthscan Publications, London.

Forsyth T and Walker A (2008) *Forest Guardians, Forest Destroyers: The Politics of Environmental Discourse in Northern Thailand.* University of Washington Press, Seattle.

Friend R, Arthur R and Keskinen M (2009) *Songs of the Doomed: The Continuing Neglect of Capture Fisheries in Hydropower Development in the Mekong.* Earthscan Publications, London.

Galer G (2004) Preparing the ground? Scenarios and political change in South Africa. *Development* **47**, 26–34.

Hajer MA (1995) *The Politics of Environmental Discourse: Ecological Modernization and the Policy Process.* Clarendon Press, Oxford.

Hazell P, Poulton C, Wiggins S and Dorward A (2007) 'The future of small farms for poverty reduction and growth'. 2020 Discussion Paper 42. International Food Policy Research Institute, Washington.

Kok K, Biggs R and Zurek M (2007) Methods for developing multiscale participatory scenarios: insights from southern Africa and Europe. *Ecology and Society* **13**, 8.

Lebel L (2006) Multi-level scenarios for exploring alternative futures for upper tributary watersheds in mainland southeast Asia. *Mountain Research and Development* **26**, 263–273.

Lebel L and Bennett E (2008) Participation in building scenarios of regional development. In *Complexity Theory for a Sustainable Future.* (Eds J Norberg and GS Cumming) pp. 207–222. Columbia University Press, New York.

Lebel L, Thongbai P, Kok K, Agard JBR, Bennett E, Biggs R, Ferreira M, Filer C, Gokhale Y, Mala W, Rumsey C, Velarde SJ, Zurek M, Blanco H, Lynam T and Yu T (2005) Subglobal scenarios. In *Ecosystems and Human Well-being. Vol. 4: Multiscale Assessment. Findings of the Subglobal Assessments Working Group.* (Ed D Capistrano) pp. 229–259. Island Press, New York.

Lebel L, Grothmann T and Siebenhuner B (2010) The role of social learning in adaptiveness: insights from water management. *International Environmental Agreements* **10**, 333–353.

Molle F, Foran T and Käkönen M (Eds) (2009) *Contested Waterscapes in the Mekong Region: Hydropower, Governance and Livelihoods.* Earthscan Publications, London.

Naurin D (2007) Why give reason? Measuring arguing and bargaining in survey research. *Swiss Political Science Review* **13**, 559–575.

Nguyen TLT, Gheewala S and Garivait S (2007) Full chain energy analysis of fuel ethanol from cassava in Thailand. *Environmental Science and Technology* **41**, 4135–4142.

Paris TR, Rola-Rubzen MF, Luis J, Truong TNC, Wongsanum C and Villanueva D (2008) Comparative analysis of the impact of labor out-migration and remittances on income and rice productivity in the Philippines, Thailand and Vietnam. Presented at AARES 53rd Annual Conference. 10–13 February 2008, Cairns, Australia.

Parnwell MJG (2006) Eco-localism and the shaping of sustainable social and natural environments in north-east Thailand. *Land Degradation and Development* **17**, 183–195.

Patel M, Kok K and Rothman DS (2007) Participatory scenario construction in land use analysis: an insight into the experiences created by stakeholder involvement in the northern Mediterranean. *Land Use Policy* **24**, 546–561.

Phongpaichit P and Baker C (1995) *Thailand: Economy and Politics*. Oxford University Press, Oxford.

Rigg J and Nattapoolwat S (2001) Embracing the global in Thailand: activism and pragmatism in an era of deagrarianization. *World Development* **29**, 945–960.

Snow DA, Rochford EB Jr, Worden SK and Benford RD (1986) Frame alignment processes, micromobilization, and movement participation. *American Sociological Review* **51**, 464–481.

Soussan J and Nilsson M (2009) *Harnessing Hydropower for Development: A Strategic Environmental Assessment for Sustainable Hydropower Development in Vietnam*. Asian Development Bank, Manila.

Thavornthon W (2009) Household expenditure, debt and happiness in the rural area. PhD thesis. National Institute of Development Administration, Bangkok.

Thongbai P, Pipattwattanakul D, Preechapanya P, Manassrisuksri K and Huaysai D (2006) Participatory scenarios for sustainable management of an ASB benchmark site in Thailand: the case of Mae Kong Kha Subwatershed of Mae Chaem Watershed.

Tomich TP, Palm CA, Velarde SJ, Geist H, Gillison A, Lebel L, Locatelli M, Mala W, van Noordwijk M, Sebastian K, Timmer D and White D (2005) Forest and agroecosystem trade-offs in the humid tropics: crosscutting assessment. Alternatives to Slash-and-Brun Consortium, conducted as a subglobal component of the Millennium Ecosystem Assessment. Alternatives to Slash-and-Burn Program, Nairobi.

UBU (Ubon Ratchatani University) (2002) Khrong kan sueksa naew tang fuen fu rabop niwet withi chiwit lae chumchon thi dai rap phon krathop chak kan sang khuean pak mun (Project to study approaches to restoration of the ecology, livelihood, and communities receiving impacts from construction of Pak Mun Dam). Ubon Ratchatani University, Ubon Ratchatani.

UNDP (2009) 'Thailand human development report 2009: human security, today and tomorrow'. UN Development Program, Bangkok.

van Notten PWF, Rotmans J, van Asselt MBA and Rothman DS (2003) An updated scenario typology. *Futures* **35**, 423–443.

Visser MP and Chermack TJ (2009) Perceptions of the relationship between scenario planning and firm performance: a qualitative study. *Futures* **41**, 581–592.

Wollenberg E, Edmunds D and Buck L (2000) Using scenarios to make decisions about the future: anticipatory learning for the adaptive co-management of community forests. *Landscape and Urban Planning* **47**, 65–77.

World Bank (2004) *Mekong Regional Water Resources Assistance Strategy: Modelled Observations on Development Scenarios in the Lower Mekong Basin*. World Bank.

World Bank and Asian Development Bank (2006) WB/ADB Joint Working Paper on future directions for water resources management in the Mekong River Basin. Mekong Water Resources Assistance Strategy. June 2006. <http://www.adb.org/water/operations/partnerships/mwras-June2006.pdf. World Bank, Bangkok>.

Index

www.ingramcontent.com/pod-product-compliance
Lightning Source LLC
LaVergne TN
LVHW060034210926
848481LV00013B/3858

* 9 7 8 0 6 4 3 1 0 4 1 2 9 *